ELEMENTARY EXPERIMENTS WITH LASERS

THE WYKEHAM SCIENCE SERIES

General Editors:

PROFESSOR SIR NEVILL MOTT, F.R.S.
Emeritus Cavendish Professor of Physics
University of Cambridge

G. R. NOAKES
Formerly Senior Physics Master
Uppingham School

The aim of the Wykeham Science Series is to broaden the outlook of the advanced high school student and introduce the undergraduate to the present state of science as a university study. Each volume seeks to reinforce the link between school and university levels. The principle author is a university professor distinguished in the field assisted by an experienced sixth-form instructor.

ELEMENTARY EXPERIMENTS
WITH LASERS

G. Wright – N.E. London Polytechnic

WYKEHAM PUBLICATIONS (LONDON) LTD
LONDON AND WINCHESTER
SPRINGER-VERLAG NEW YORK INC.
1973

Sole Distributors for the Western Hemisphere
SPRINGER-VERLAG NEW YORK INC. NEW YORK

Cover illustration – Transmitted and reflected orders of reinforcement for laser beam incident on grating from the left.

ISBN 0-387-91063-8

Library of Congress Catalog Card Number 70-135384

First published 1973 by Wykeham Publications (London) Ltd.

© 1973 G. Wright. All rights reserved. No part of this publication may be reproduced, stored in a retrieval system, or transmitted, in any form or by any means, electronic, mechanical, photocopying, recording or otherwise, without the prior permission of the copyright owner.

Printed in Great Britain by Taylor & Francis Ltd.
10–14 Macklin Street, London, WC2B 5NF

PREFACE

NOWADAYS books on lasers abound, but there is an evident gap between the 'Science for the Layman' type of book and the more advanced works aimed at final year students and above. The present book is an attempt to fill this gap, and to do so in two ways.

Chapters 1–4 are concerned with the development of some experiments that can be done with a simple gas laser, which is the commonest laser used in schools. The list of experiments is not by any means exhaustive and is largely based on traditional ways of teaching optics. The laser does make it possible to approach the development of optics in different ways, but for reasons of space I have stayed with the traditional views as set out in established optics text books.

Chapters 5–8 deal with the properties and construction of lasers. This section is not confined to gas lasers since it is designed to answer the broader questions that arise when a student first uses a laser.

Those who will use lasers in the lecture theatre or laboratory, should read the appendix on safety very carefully. A knowledge of the material referred to in this section is essential for the successful use of lasers in educational establishments.

I must express my sincere gratitude to Norman Weeden who took all the photographs in the text, without which the book would be much the poorer. I would also like to thank my schoolmaster colleague Geoffrey Foxcroft for his many helpful suggestions during the preparation of this book.

I gratefully acknowledge my indebtedness to the Experimental Notes of Messrs. Griffiths and George for the material of pages 21 and 22.

G. WRIGHT

CONTENTS

Preface		v
Chapter 1	INTRODUCTION	1
Chapter 2	HOLOGRAPHY	35
Chapter 3	GEOMETRICAL OPTICS	55
Chapter 4	POLARIZATION	64
Chapter 5	LIGHT EMISSION AND ENERGY LEVELS	74
Chapter 6	COHERENCE	86
Chapter 7	RESONANCE AND LINE WIDTH	98
Chapter 8	LASER CONSTRUCTION	107
Appendix	SAFETY	123
Index		127

CHAPTER 1
introduction

THIS book was written to show many of the quite well-known demonstrations in physical optics as they can be done with a gas laser as the light source. The early chapters concentrate on this practical side. The later part explains in reasonably simple terms how lasers themselves work and why they have such useful properties; it also goes rather more deeply into some questions, such as line width and coherence, which lasers either raise or help to answer.

The three features of the monochromatic emission from a laser that make it so much more useful as a source for our purposes than, say, a sodium discharge lamp, are: greater coherence (a term to be explained later), the nearly parallel beam, and the high luminous flux per unit area in the beam.

Many of the experiments described in this first chapter can be done in a limited way with discharge lamps or similar sources, but a laser enables a large group of people to see the same demonstration together, and in more detail.

1.1. Single slit diffraction

The simplest procedure (fig. 1.1) is to place a single slit in the path of the laser beam and observe the diffraction pattern on an opaque screen placed beyond the slit. The Fraunhofer diffraction pattern is a series of blobs of maximum brightness (fig. 1.2) approximately equally

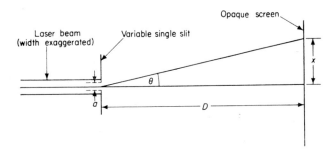

Fig. 1.1. Arrangement for single-slit diffraction.

spaced. The central maximum is twice as wide as the others. The brightness *minima* between these maxima occur at angles θ given by $\sin \theta = p\lambda/a$, where p is an integer (1, 2, 3, . . .), λ is the wavelength and a the width of the slit aperture. If D is about 1 m and a about 0·1 mm the pattern spreads over a width of about 50 mm; then θ is so small that $\sin \theta$ (and $\tan \theta$) both approximate to θ, and the minima are very nearly equally spaced.

A variable-width slit enables us to study the effect of a. With a relatively wide slit, at least 20 maxima on either side of the centre can be observed with a 0·25 mW helium-neon laser beam. As the slit is narrowed, individual maxima spread out (fig. 1.3) until the central maximum fills the whole field (fig. 1.4). With D equal to several metres, the distance of the pth minimum from the centre can be measured with a metre rule, $\tan \theta$ and hence λ/a calculated, so, if a is measured with a travelling microscope λ, the wavelength of the laser light, can be found.

A result more like traditional single-slit patterns is obtained if the narrow laser beam is expanded so that it fills the whole length of the

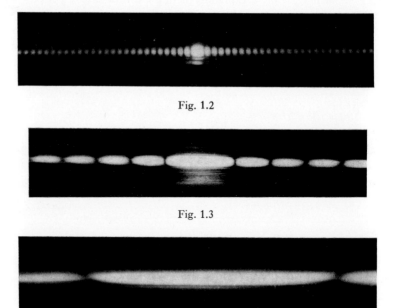

Fig. 1.2

Fig. 1.3

Fig. 1.4

Figs. 1.2 (widest slit), 1.3 and 1.4 (narrowest slit). Fraunhofer diffraction patterns with direct laser beam; slits of different width. Note the wide central maximum in each case.

slit, using a short-focus cylindrical lens (focal length between 50 and 100 mm), which converts a beam of circular cross-section into one of elliptical cross-section. The pattern has the same lateral distribution as in the previous experiment (compare figs. 1.5, 1.6, 1.7 and figs. 1.2, 1.3, 1.4) but we get lines instead of blobs. Fewer maxima are observable, as spreading the light out reduces the brightness of the pattern. (The tapering in fig. 1.7 is because, when the slit is very narrow, slight variations in the width have a large proportional effect on an already small width. Also, the horizontal stripes are due to imperfections of the slit edges.)

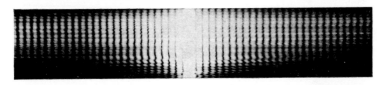

Fig 1.5

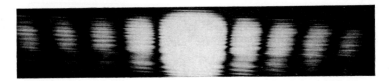

Fig. 1.6

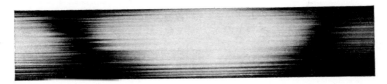

Fig. 1.7

Figs. 1.5 (widest slit), 1.6 and 1.7 (narrowest slit). Fraunhofer diffraction patterns with vertically expanded laser beam, corresponding to Figs. 1.2, 1.3, 1.4.

The intensity of the laser beam makes it possible to demonstrate single-slit diffraction in three dimensions in a smoke box which is held close to the slit—or which has the slit mounted inside it (fig. 1.8). (A suitable smoke box and smoke generator are described in the Nuffield Physics *Guide to Experiments*, Year III.)

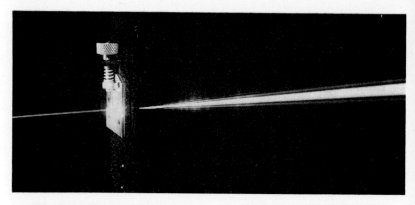

Fig. 1.8. Laser beam emerging from a single slit in a smoke box.

1.2. *Rectangular, square and circular apertures*

With two narrow slits at right angles to each other, each produces its own diffraction pattern. Thus, in fig. 1.9, slit AB gives a pattern parallel to XY, while slit XY gives a pattern parallel to AB. A rectangular aperture is equivalent to a long narrow slit, with a short fat slit at right angles to it. It is simpler to replace such an arrangement of slits at right angles to each other by a rectangular hole or aperture which will produce an identical diffraction pattern (fig. 1.10). Note that the *larger* distance between minima is due to diffraction across the *narrower* width of the aperture.

A simple method of making rectangular apertures is to lay parallel strips of heavy-duty black masking tape on a microscope slide, so that their edges almost touch each other (this will also work as a fixed single slit). Then another pair of strips are added at right angles to the first

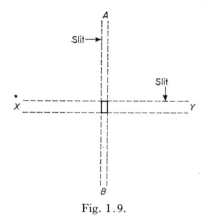

Fig. 1.9.

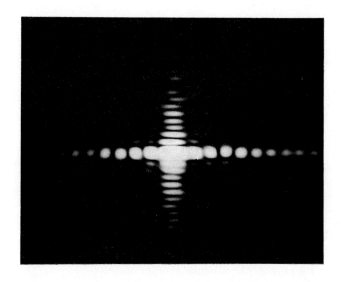

Fig. 1.10. Diffraction pattern from rectangular aperture.

pair, which leaves in the centre of the slide a small rectangular aperture. Suitable approximate dimensions of the aperture are 2 mm × 1 mm.

More accurate apertures may be made by photographic reproduction. The aperture sizes required are scaled up by a suitable factor, such as 20 times, to enable appropriate circles, squares and rectangles to be drawn in Indian ink on white card. These are then photographed from some distance away, for example, from the other side of a laboratory. A plate camera is used. The negative obtained in this way will then have the required shape apertures on it, and the distance will provide a demagnifying factor of about 20 times. Any high contrast line emulsion plates will be suitable since such plates have very few grey shades between the black and white extremes of exposed and unexposed. This is a fairly critical process, especially at the developing stage. If the plate is under-exposed or under-developed, light will pass through the areas surrounding the aperture; and, if it is over-exposed the boundary is not always sharply defined. Test exposures are essential in this type of work.

Figures 1.11–1.13 show the diffraction patterns produced by circular apertures of varying radius, 0·3 mm, 0·6 mm and 0·9 mm, respectively. The experimental arrangement necessary to produce these patterns is the same as that used for the dotted single-slit diffraction pattern. The shape of this pattern is known as the Airy disc, after its first observer, who, with the apparatus available at the time was able to see little more than the first and second maxima, and the first minimum, which occurs at an angle θ given by the equation

$$\sin \theta = 1\cdot 22 \lambda/a,$$

where a is the diameter of the aperture, and λ the wavelength of the laser radiation. The other minima are not linearly spaced, in contrast to the minima for the single-slit pattern.

Fig. 1.11. Circular aperture, radius 0·3 mm.

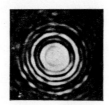

Fig. 1.12. Circular aperture, radius 0·6 mm.

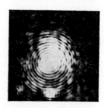

Fig. 1.13. Circular aperture, radius 0·9 mm.

1.3. *Limit of resolution and its measurement*

One of the most fundamental concepts in the design of optical instruments is diffraction as applied to image formation and resolution.

In fig. 1.14, light from a point source A strikes a screen S which has a small circular aperture; the light passing through the aperture is focused on a screen T by a lens which is much wider than the aperture, and a diffraction pattern of the type shown in fig. 1.11 is observed on T where we might expect just to find a point image of A. Now, this happens

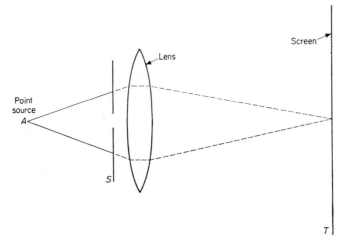

Fig. 1.14.

whenever we think we are using a lens to form a point image of a point object—but it is usually inconspicuous, partly because the pattern is small if the aperture is reasonably large, and partly because it is often swamped by inevitable lens defects.

The image of two bright object points that are close together consists of two sets of concentric rings, one for each source (fig. 1.15(a)), and here they are identifiable as separate patterns, since the central maxima are distinguishable. If the object points are closer together, the pattern is as fig. 1.15(b). There must be a limit to the closeness at which objects can be distinguished in the image, and this is the stage at which it is just possible to see that there are two separate central discs. This limit is the *limit of resolution* for the optical system. For a microscope it is measured as the smallest *distance* between object points, such that the central maxima of their diffraction patterns are separable; for the

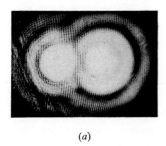

(a)

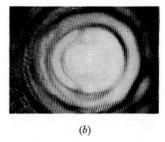

(b)

Fig. 1.15. Overlapping diffraction patterns from two apertures, separable in (a) but not in (b).

telescope it is the smallest *angle* between two distant object points, such that the central maxima of their diffraction patterns are separable. The decision whether two images are separable or not is a subjective one, and to limit uncertainty, Rayleigh proposed his criterion for resolving two point objects which is that the middle of the central maximum of the one diffraction pattern coincides with the first minimum of the other diffraction pattern.

To investigate this, a circular aperture, about 0·25 mm diameter, is made photographically (p. 5) in the centre of a plate. A line AB is ruled on the plate with a razor blade, within 2 mm of the aperture, and the emulsion on the far side of AB is scraped off (fig. 1.16 (*a*)). This plate may be mounted in a holder on an optical bench, with freedom to rotate about an axis parallel to the bench. A second similar plate is mounted in another holder, the two are moved as close together as the mountings allow, and they are adjusted so that their opaque areas overlap (fig. 1.16 (*b*) which is not to scale) when the light transmitted through the hole in one plate is transmitted through the clear part of the other. The narrow beam from the laser is widened by a telescopic system (fig. 1.17) so that it can cover both holes.

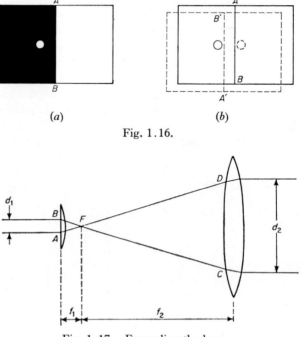

Fig. 1.16.

Fig. 1.17. Expanding the beam.

Figure 1.15 shows what is observed on a screen 1 m away. If the screen is moved further away from the holes the pattern changes from type (*a*) to type (*b*). These pictures show not only the two sets of circular diffraction rings, but also parallel fringes of the ' Young's slits ' type which are discussed on page 10.

If an accurately parallel beam is used, and this may be checked by measuring the diameter of the widened beam at two well separated points, the spacing of the holes is also the spacing of the centres of the two components of the composite diffraction pattern. A wooden rule provides a rough measure of the spacing of the holes. If the holes are illuminated by a **non**-laser source, then an accurate measurement of their separation may be obtained using a travelling microscope. A microscope must **not** be used with a laser beam. The screen is moved until the two patterns are just resolved according to the Rayleigh criterion, and the distance D from the holes to the screen is measured with a metre rule. For small angles $\sin \theta = \tan \theta = x/D$. Now, $\sin \theta = 1 \cdot 22 \lambda/a$ (p. 6), so $1 \cdot 22 \lambda/a$ should equal x/D. Different observers will probably estimate D differently!

A refinement is to mount the first plate as before and the second in a holder which allows both vertical and horizontal movement. AB and A'B' are made parallel to whichever of these two movements is the finer. When this fine adjustment is made the separation of the two holes alters and the change in diffraction pattern may be observed, which helps to emphasize the idea of resolution.

When students first meet the topic of resolution, they may not appreciate its relevance to optical instruments, since small apertures (1 mm diameter or less) are needed to demonstrate these diffraction patterns, and one's early experiences with lenses deal with lenses 20 mm diameter or more. Examination of a $\times 40$ or $\times 100$ microscope objective (fig. 1.18) shows that diffraction in such cases is indeed

Fig. 1.18. A $\times 40$ objective; the scale reads centimetres.

relevant. When such a lens has been placed in a laser beam, a pattern similar to fig. 1.13 appears on a screen 2 m away; the distance of the screen has been used to magnify the pattern, but it definitely exists on a smaller scale in the image plane when the objective is used normally. (Unfortunately, because of the high coherence and intensity of the laser beam, other diffraction patterns arise, due to multiple reflections in the complex lens; these reflections provide other beams, equivalent to additional coherent sources, which in turn produce further diffraction patterns. Also, dust particles on the lens surfaces produce yet other diffraction patterns.)

1.4. *Young's slits*

For this demonstration in its simplest form, we need only a gas laser, a fogged photographic plate (or a clear glass plate coated with 'Aquadag'), an opaque screen, a razor blade and a steel rule. Two lines about 0·2 mm apart are ruled on the plate, which is then placed in the beam between the laser and the screen, where a pattern of regularly spaced blobs (fig. 1.19) will appear. An essential difference between this and the pattern due to diffraction by a single slit is that the central maximum is the same width as the other maxima, while in the diffraction pattern the central maximum is twice as wide as the rest. It is useful to remember this point when doing physical optics experiments with a laser, since, because of the high coherence of the laser beam, interference and diffraction patterns are obtained so easily that it is possible to get both types of pattern together when we are really interested in emphasizing one type only.

If the screen is replaced by a metre rule about 10 m from the slits, the fringe separation is about 3 cm, which is easily measurable on the rule. The average spacing between adjacent fringes $(x_n - x_{n-1})$ may be found. The distance D from slits to screen may also be measured by a metre rule. The spacing d between the two slits can be measured with a travelling microscope, when the wavelength, λ, of the light is calculated from the equation $(x_n - x_{n-1}) = \lambda D/d$.

The slits illuminated by the narrow laser beam are effectively *point* sources (as in Young's original experiment) and the beam has to be

Fig. 1.19. 'Two-slit' interference patterns with direct laser beam.

expanded if it is to illuminate the whole length of the slits. Most theoretical treatments of the experiment actually deal with point sources, never explaining the extrapolation that the simple theory is true for slits as well as for points. Thus, as here described, the experiment is closer to the usual presentation of the theory than is the conventional arrangement. If the beam is expanded, as described on pp. 2 and 3, so as to cover the whole length of the slits, then the pattern with parallel fringes instead of spots is obtained. (Figs. 1.20 (*a*), (*b*), (*c*) which are for different spacings the slits are furthest apart for (*a*) and closest for (*c*).)

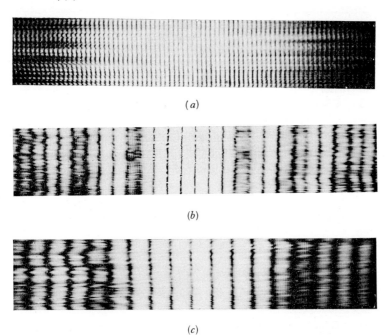

Fig. 1.20. 'Two-slit' interference patterns with expanded beam.

1.5. *Lloyd's single mirror*

It is possible to obtain two images of a single point and to use these two images as two sources for a 'double-slit' experiment. The laser considerably simplifies such systems. For Lloyd's single mirror, a converging lens (*f* about 100 mm) is used to convert the parallel laser beam into a point source S at a finite distance (fig. 1.21). A polished glass block (as used in simple optics experiments) is introduced from the side of the laser beam until it reflects part of the divergent beam and this produces a virtual image of S at S_1. These two point

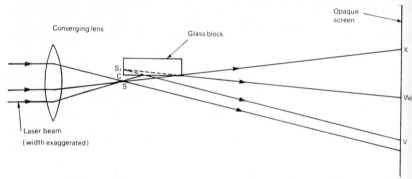

Fig. 1.21. Arrangement for Lloyd's single mirror.

sources provide a two-slit interference pattern on an opaque screen placed some distance away, about 5 m.

The corner C of the block should be very close to or at the focus of the lens, since otherwise the separation of S and S_1 will be so great that very fine, barely visible fringes or invisible fringes will be formed. This adjustment may be difficult to judge if the diameter of the laser beam, just before reaching the converging lens, is small (about 2 mm). This diameter may be enlarged by either placing the lens some distance from the laser (1 m or more) and using the natural divergence of the beam, or by placing a second converging lens (focal length about 100 mm), as in fig. 1.22, where the width of the laser beam has been exaggerated.

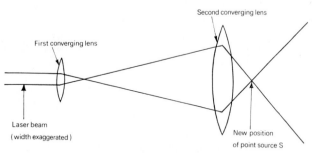

Fig. 1.22. Improving definition of a point source.

As the block is introduced two beams of light will be observed on the opaque viewing screen. The block position is adjusted, both twist and sideways movement being used, until the two beams, one from S and the other from S_1, overlap. Within the region VW (fig. 1.21) the evenly spaced 'double-slit' fringes may be observed. On the extreme

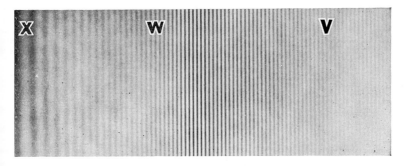

Fig. 1.23. Lloyd's single mirror fringes.

edge XW of the beams a second pattern (fig. 1.23) will appear. This is a diffraction pattern due to the straight edge of the block (for further details of 'straight-edge' patterns see pp. 24, 27). The basic equation for this experiment is the same as that used in the Young's slits experiment (p. 10), as are all the measurements *except* for the measurement of d where a travelling microscope must **not** be used.

In this experiment d may be measured by a magnification method. A converging lens (focal length about 100 mm) is introduced between S and the opaque screen. The position of the lens is adjusted until images of S and S_1 are focused on the screen. Their separation is measured with a ruler and the ratio of object to image separation equals the ratio of object to image distance. This provides a means of measuring d. There are two possible positions of the lens for image formation on the screen. The one with the shorter object distance must be used if magnification is to be obtained.

1.6. *Fresnel's biprism*

The basic laser properties enable this experiment to be done in a simpler way than the traditional arrangement which has a single slit preceding the biprism. As previously described for Lloyd's single mirror experiment (p. 11) a converging lens (focal length 100 mm) is used to convert the parallel laser beam from an apparent point source at infinity to a finite point source, in this case about 40 mm in front of the biprism. The biprism converts this point source into two virtual images which are the basis for a 'double-slit' pattern (fig. 1.24). The point focus has the advantage over a single slit in that all the available light energy passes through the biprism, whereas the single slit blocks out a high proportion of the beam. In addition the single slit provides a diffraction pattern superimposed on the basic interference pattern (p. 17). This factor is eliminated by this simpler method.

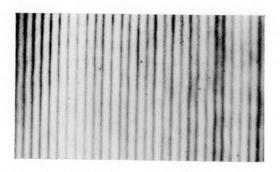

Fig. 1.24. Biprism fringes.

1.7. Equal amplitude condition

It is assumed in the theory for Young's slits and similar experiments that the amplitudes of the two sets of waves that are superposed to give rise to the pattern are equal. This can be investigated in terms of intensity I, which is proportional to the *square* of the amplitude A; $I \propto A^2$, and $A \propto \sqrt{I}$.

With the help of a light meter we can measure the attenuated intensity of a light beam after it has passed through a stack of microscopic slides; if this is reduced to a tenth, the amplitude has been reduced by a factor $1/\sqrt{10} \approx 0{\cdot}32$. We can then use such stacks to cover one of the slits, and see the effect on the contrast.

1.8. The influence of diffraction on two-slit interference

The very elementary theory of Young's slits really assumes that the slits are so fine that the whole field is an overlapping of the central diffraction maxima from the two slits. This never happens with slits that are wide enough to be used. Each slit on its own would produce its own pattern of diffraction maxima and minima, and as the slits are very close together the two diffraction patterns overlap and are effectively a single one. But the spacing of the diffraction maxima and minima is wider than that of the interference fringes, which appear with a spacing governed by the two-slit interference condition and an intensity (in their own maxima) governed by the diffraction pattern.

Figure 1.25 is in three parts, and relates to a two-slit Young's fringes arrangement, with the screen about 5 m from the slits. The top (*a*) shows the diffraction pattern of one slit on its own, the bottom (*c*) that for the other slit on its own, and the middle (*b*) shows the effect with both slits in action. They were all taken on one plate, and the slits masked in sections to give the corresponding pictures (fig. 1.26).

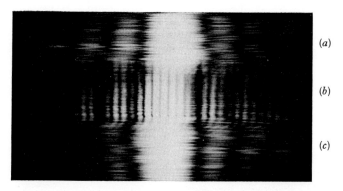

Fig. 1.25. Young's fringes (b), together with (a) and (c) the diffraction pattern of the individual slits.

As with Young's slits, when a laser is used, beautiful and extremely bright fringes are obtained with a biprism without a primary slit; and for introductory work this procedure is obviously the best. However, some interesting observations may be obtained if a primary slit is used with a biprism on a conventional optical bench—although it should be emphasized that these refinements are not essential for a simple demonstration of the interference pattern itself.

This experiment is worth additional discussion, since its patterns can turn out to be more complex than those of fig. 1.25 when the traditional single slit arrangement of fig. 1.27 is used, and the complexities may be traced to particular physical situations. To observe these changes in the interference and diffraction patterns the spacing of the components is altered. This is best achieved by having a single slit and the biprism mounted in holders capable of fine transverse adjustment across the laser beam. This is more important than making the apex of the prism accurately parallel to the single slit. The two halves of the biprism form two virtual images of the slit in the same plane as the

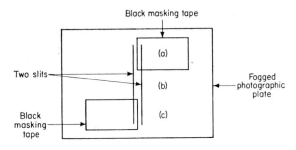

Fig. 1.26.

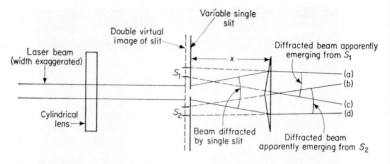

Fig. 1.27. Traditional biprism arrangement. The virtual-image plane and the plane of the slit have been separated for the sake of clarity.

slit itself—a 'virtual-image double slit' in effect. (If these 'virtual-image slits' are to be viewed directly, then we *must* replace the laser with a non-laser source such as a tungsten filament lamp.) The greater the distance x between the object slit and the biprism, the further apart are the 'virtual-image slits'. Alteration of x provides a very convenient demonstration of the change of spacing of the interference fringes with the change of separation of the slits. If the single slit is narrow and the distance x is fairly small, say, 5 cm, then the pattern obtained is the simple interference pattern of fig. 1.24.

If the separation x is increased the fringes become closer together and more of them appear but confined to the same total width as before. Any envelope or modulation pattern present becomes more prominent. When x is about 10 cm, the single-slit diffraction pattern is clearly seen. That it is really the single-slit diffraction pattern may be demonstrated by altering the width of the slit itself, when there is a corresponding change in the envelope pattern, and the spacing of the fringes remains unaltered. Figures 1.28 (*a*), (*b*) and (*c*) are for increasing slit widths. When the single slit is very narrow, (fig. 1.28 (*a*) the central (zero order) diffraction maximum covers the whole field and the interference fringes appear to be of uniform contrast.

If x is increased still further, say to 20 cm, an additional pattern is superimposed as shown in fig. 1.29, and this is due to diffraction at the straight edge of the apex of the prism. Figure 1.27 is not drawn to scale, and shows the relative positions of the two images of the slit. The apex of the biprism, the angle of which is about 179° has been considerably exaggerated, and four specific positions (*a*), (*b*), (*c*) and (*d*) in the field of the pattern are shown. Slit 1, the prism apex, and position (*d*) are nearly line, as are slit 2, the apex of the prism, and position (*a*). This additional superimposed

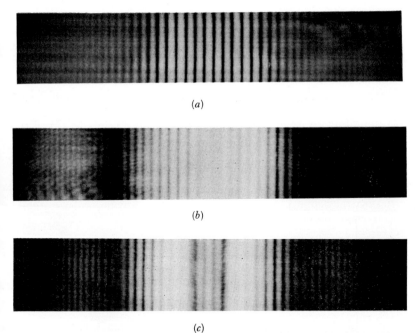

Fig. 1.28. Biprism fringes: slit-widths $(a) < (b) < (c)$.

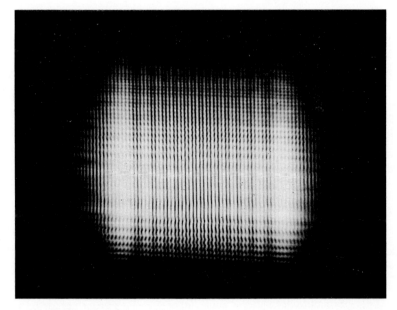

Fig. 1.29. Biprism fringes; very narrow slit. This pattern appears to take a different form when viewed from a distance.

pattern starts at the geometrical shadow due to the straight edge of the apex of the prism, but because of diffraction effects this edge is not sharply defined. This may be shown even more clearly by a divided biprism experiment, rather like the divided Young's slits experiment described on p. 15. The biprism is masked with black heavy-duty masking tape as shown in fig. 1.30. Figure 1.31 shows the pattern corresponding to sections (*a*), (*b*) and (*c*) of the biprism; (*b*) is the full pattern, (*a*) is the straight edge diffraction pattern due to S_1, and (*c*) is the straight-edge pattern due to S_2. These complex patterns raise a question which is difficult for students. What is the difference between an interference pattern and a diffraction pattern, since they appear to be observed in the same way in similar systems?

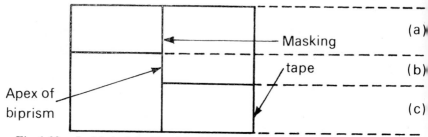

Fig. 1.30. In (*a*) the left side of the biprism is masked, and in (*c*) the right side; in (*b*) there is no masking.

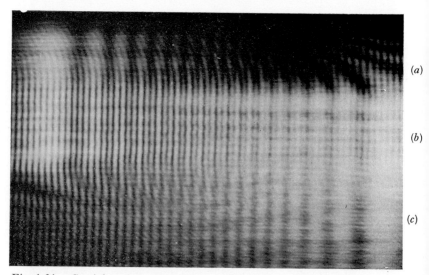

Fig. 1.31. Straight-edge patterns alone, (*a*) and (*c*) due to edge of biprism and respectively S_1 and S_2. In (*b*) there are both of these patterns together, together with an interference pattern similar to that of fig. 1.29.

Interference at a general point in the plane of observation is due to the superposition of two or more wavefronts, each separated by a finite path difference; for example, light from two very thin slits separated by a finite distance.

Diffraction takes place when the effect at a general point in the plane of observation is due to the sum of the contributions from an infinitely large number of small neighbouring elements of a wavefront, with infinitesimally small path differences between them.

For example, the slit in the single-slit diffraction experiment may be considered to be made up of a very large number of infinitesimally small sources in a row, the contributions for each having slightly different paths to the observation plane.

The form of the variation of the path difference between neighbouring elements is used to define two classes of diffraction effect: *Fraunhofer diffraction*, in which the path difference between elements is directly proportional to the separation of the elements, and *Fresnel diffraction*, in which this variation may be any mathematical function more complex than the direct proportion of Fraunhofer diffraction.

1.9. Newton's rings

The interference experiments described so far are all based on 'division-of-wavefront'. Two or more sections from separate parts of the wavefront are made to follow different optical paths until they eventually overlap in the plane of observation.

There is another type of interference based on 'division-of-amplitude'. In this case the energy of the radiation that makes up a section of a wavefront is divided into two parts; for example, by striking a glass surface, in which case part is transmitted and part is reflected. When these two parts are later superimposed they may interfere. This process applies for each individual section of the wavefront and hence for the whole of it. Newton's rings, as usually observed, are formed with a long-focus converging lens (and indeed described by Newton) on a plane glass surface. But the lens can be used alone.

An undeviated laser beam (fig. 1.32) is projected on to a safety screen (K) some distance (about 5 m) from the laser. A long (about 1 m) focus lens (J), which provides the division of amplitude, is placed in the laser beam and close to the safety screen. An opaque viewing screen is placed on the laser side of " Newton's rings lens " close to the laser but not blocking the laser beam. Reflections from each of the surfaces of the lens produce two patches of light on the viewing screen. Careful horizontal and vertical movement of the lens will produce overlap of the two patches of light at the viewing screen and within

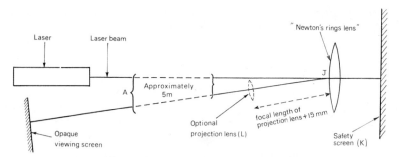

Fig. 1.32. Newton's rings apparatus.

this patch will be a small interference ring pattern. Movement of the viewing screen along the overlapping light paths shows that the interference takes place in all planes where the beams overlap.

This pattern may be magnified by placing a projection lens (a converging lens of focal length about 100 mm), so that the separation of the lenses is about 15 mm greater than the focal length of this projection lens. The viewing screen is placed as far away from the projection lens as is conveniently possible. A magnified interference pattern will appear on the viewing screen. The magnification factor will be the ratio of image distance to object distance for the projection lens.

It was mentioned earlier that " Newton's rings lens " formed interference patterns in all planes where overlap occurred. These patterns have different sizes in different planes. If measurements are to be made it is essential to know which of these many patterns has been projected on to the screen. A convenient method is to switch off the laser and illuminate " Newton's rings lens " with a white light source. The projection lens is adjusted until a clear image of " Newton's rings lens " is formed on the viewing screen. When the laser is switched back on for the projection lens in this setting the pattern at the lens surface is the interference pattern projected on to the screen.

In the traditional Newton's rings experiment, with parallel light incident normally on the system and the pattern viewed in reflected light, the reflections from the under surface of the lens and the upper surface of the optical flat are of comparable amplitude; but one has undergone a phase change of π and the other has not, so that an equivalent path difference of $\lambda/2$ has to be added to the geometrical path difference. At a distance r from the centre of the pattern the geometrical path difference is $2t$, where t is the thickness of the air film. If the optical path difference is a whole number of wavelengths, then reinforcement occurs. The centre of the pattern is dark and for the first bright ring $2t = \frac{1}{2}\lambda$, for the next $2t = 1\frac{1}{2}\lambda$, for the next $2t = 2\frac{1}{2}\lambda$ and so on (if contact at the centre is perfect.)

Light from a laser is so highly coherent (and therefore highly monochromatic) that with a laser source quite large path differences may be used—such as the thickness of a " Newton's rings lens ". Hence the optical flat of the conventional arrangement is not needed when a laser is used. In fact an optical flat is a positive disadvantage in that interference will take place between every pair of surfaces in such an arrangement. In this case the optical path difference at thickness t is $2nt$, where t is the thickness of the lens and n the refractive index of the glass.

1.10. *Measurement of wavelength using Newton's rings*

An image of the rings formed on the front surface of the " Newton's rings lens " at J is projected on to the screen at K (fig. 1.32) and the magnification is found by measuring the object and image distances for the projection lens at L. This enables the dimensions of the rings at the lens surface to be deduced, and from this, the wavelength λ can be obtained. If the maximum lens thickness is D, then the thickness, t, at a point r from the centre is

$$t = \left[D - \frac{r^2}{2}\left(\frac{1}{S_1} + \frac{1}{S_2}\right)\right]$$

where S_1 and S_2 are the radii of curvature of the lens surfaces (fig. 1.33).

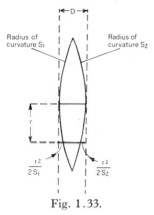

Fig. 1.33.

Ignoring the effect of obliquity, which introduces quite a small error, the optical path difference between light reflected at the front and back faces is then

$$2n\left[D - \frac{r^2}{2}\left(\frac{1}{S_1} + \frac{1}{S_2}\right)\right] \pm \lambda/2,$$

where n is the refractive index of the glass. Thus, because of the phase change accompanying reflection at the first surface, the condition for *darkness* is

$$2n\left[D-\frac{r^2}{2}\left(\frac{1}{S_1}+\frac{1}{S_2}\right)\right]=p\lambda,$$

where p is an integer.

For successive rings, p changes by 1, decreasing as the ring radius increases. For the xth ring from the centre, radius r_x,

$$2nD-r_x^2 n\left(\frac{1}{S_1}+\frac{1}{S_2}\right)=(p-x)\lambda,$$

where p is now the order of interference at the centre. The ring radius, R_x, measured on the screen is m times larger than r_x, where m is the magnification produced by the short focus lens.

$$\therefore\ 2nD-\frac{R_x^2 n}{m^2}\left(\frac{1}{S_1}+\frac{1}{S_2}\right)=(p-x)\lambda.$$

Thus the gradient of a graph of R_x^2 against x is $\dfrac{m^2\lambda}{n}\left(\dfrac{1}{S_1}+\dfrac{1}{S_2}\right)$. The values of n, S_1 and S_2 may be found by conventional methods.

When a conventional Newton's rings apparatus is illuminated by laser light, interference will take place between light reflected from four surfaces, the two surfaces of the lens, and the two surfaces of the flat; the resulting pattern will be complex.

So with laser illumination we can simplify the system and use only the two surfaces of the lens for division of amplitude, leaving out all the rest of the conventional apparatus. Normally contour fringes require a microscope, but (for obvious safety reasons) microscopes must **not** be used in conjunction with laser illumination. Projection of the fringes on to a distant screen is an alternative method of providing magnification, and the laser is sufficiently intense for this to be done successfully.

Other division-of-amplitude systems may be used in the same apparatus instead of the long-focus lens. Suitable examples are parallel faces of a good quality glass block and microscope slides. This experimental arrangement may be used to check the flatness and parallel nature of the surfaces of microscope slides in order to find suitable optical flats.

1.11. *Fresnel diffraction*

In the single-slit diffraction experiment (p. 1) the laser, because of its nearly parallel beam, provides the equivalent of a point light source at infinity. The viewing screen was approximately 1 m or more away from a slit of width 0·1 mm, which may be considered as being equivalent to viewing the diffraction pattern at infinity. This is an example of Fraunhofer diffraction.

In simple terms Fresnel diffraction occurs when the source or screen or both occur at a finite distance from the diffracting aperture or obstacle. This is the converse of the point source and screen at infinity that applies to Fraunhofer diffraction.

A converging lens (focal length about 200 mm) is placed in the path of a laser beam and a slit placed 300 mm away from the lens. The lens focuses the laser beam to a point as in fig. 1.34. This point acts as a point source a finite distance a from the slit. An opaque viewing screen placed on the other side at a distance b from the slit will display a diffraction pattern of the type shown in figs. 1.35 and 1.36. These patterns change as the distances a and b change and as the slit width alters. The change of pattern contains two elements. One is a change of relative distribution of the maximum and minimum positions of brightness and the other is a change of size. This change of size cannot be a precise and constant factor since the distribution of bright and dark lines also varies. The distance between a maximum and its neighbouring minimum is very approximately proportional to the form $\sqrt{\{b(a+b)/a\}}$, which is useful as a guideline when arranging Fresnel

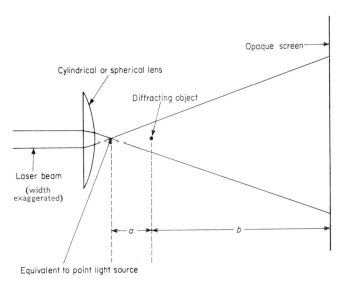

Fig. 1.34. Arrangement for Fresnel diffraction.

diffraction. With a and b both 1 m and for light of the visible spectrum, the approximate separation of neighbouring maxima is 1 mm.

In order to obtain larger patterns to demonstrate to larger groups, either the distance b is made larger or the distance a is made smaller or both. To be of significance a needs to be made about 100 mm or b made greater than about 5 m. The latter arrangement is more suitable because larger objects can be placed in the beam. The distance a from the light source to the object is the distance from the object to the principal focus of the lens and not the distance to the lens itself.

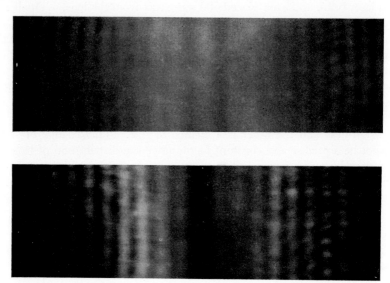

Figs. 1.35 and 1.36.   Fresnel diffraction pattern, single slit.

Typical Fresnel diffraction patterns are shown in fig. 1.37 (a), (b) and (c). For long objects such as the pin a cylindrical lens is more suitable since its cylindrical wavefronts concentrate the light energy in the region of the object. A spherical lens producing spherical wavefronts is more suitable for square or round objects such as razor blades or ball bearings. The object must be placed so that it is surrounded by overlap from the diverging beam, because it is this overlap that provides the wavefronts that produce the diffraction pattern.

The straight-edge pattern is an exceptional case of Fresnel diffraction in which the general appearance of the pattern is independent of the 'object' and 'image' distances a and b. This pattern was also shown in Fresnel's biprism experiment and Lloyd's single mirror experiment (pp. 13, 17 and 18 and figs. 1.23, 1.29 and 1.31).

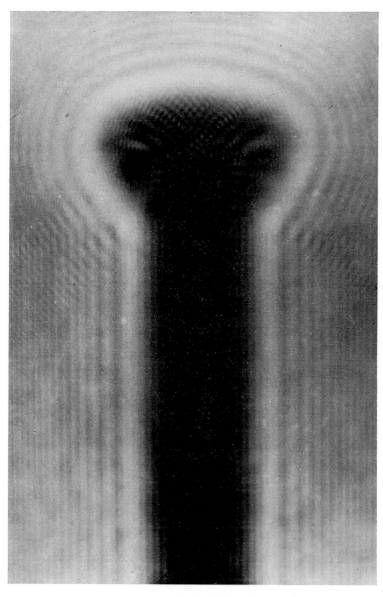

Figure 1.37 (*a*). Fresnel diffraction pattern.

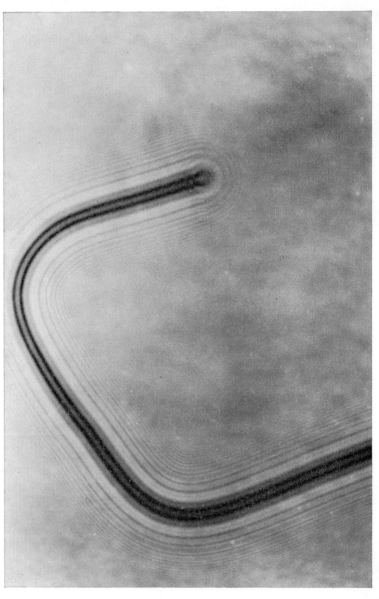

Figure 1.37 (*b*). Fresnel diffraction pattern.

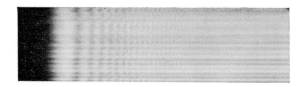

Fig. 1.37 (c). Fresnel diffraction pattern, 'straight-edge'.

1.12. *Diffraction grating*

According to the distinction drawn on p. 19, the diffraction grating ought to be called the interference grating, since the summation by the principle of superposition that provides the basic equation for the position of the different order spectra is a summation of finite differences of phase. However, historically, it was called the diffraction grating and diffraction processes play a major part in the development of the theory of resolution of the grating.

This basic equation is, for normal incidence,

$$d \sin \theta = p\lambda,$$

where d is the spacing of the lines on the grating, θ is the angle of observation relative to the grating normal, p is the order of the diffraction, i.e. the number of wavelengths path difference between the contributions of neighbouring slits in the grating, λ is the wavelength of the monochromatic light that is incident normally on the grating. $p = 0$ for the central 'white' maximum; when $p = 1, 2$, the spectra lie on either side of the normal to the grating.

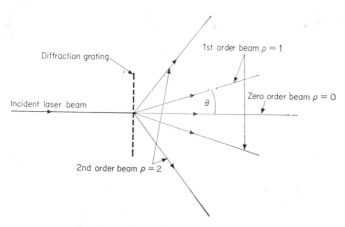

Fig. 1.38. Diffraction grating.

One of the most fascinating demonstrations is to put a diffraction grating into a laser beam, and obtain the diffraction maxima on the ceilings and walls of the room. If this is done with a grating with 6×10^5 lines per metre ($d = 1\cdot 67 \times 10^{-6}$ m), then two orders will be seen (fig. 1.38). The third order is not obtained with the red wavelength of the helium–neon laser. One can also observe reflected maxima at angles which (for the same λ and for normal incidence), are the same numerically as those for the transmitted orders (fig. 1.39).

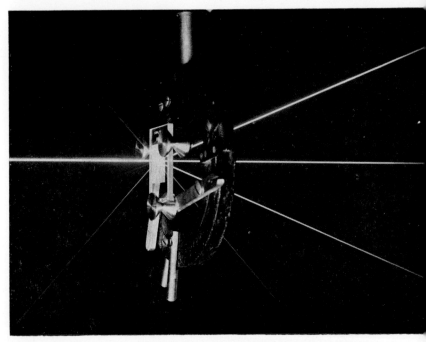

Fig. 1.39. Transmitted and reflected orders of reinforcement for laser beam incident on grating from the left.

If a grating with this spacing is used and the beam is not incident normally as in fig. 1.40, then higher orders may be obtained. For an angle of incidence i,

$$d (\sin i - \sin \theta) = p\lambda.$$

One needs to watch signs here; for example, if θ is measured from the normal in fig. 1.40 then it is negative when measured clockwise and positive when measured anticlockwise. The maximum values for i and θ are 90°, so that the limiting value of p is given by

$$p\lambda = d\{1-(-1)\} = 2d = \frac{2}{6 \times 10^5} \text{ m.}$$

28

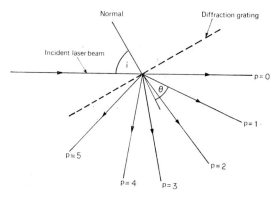

Fig. 1.40. Non-normal incidence.

So

$$p = \frac{2d}{\lambda} = \frac{2}{6 \times 10^5 \times 6 \cdot 33 \times 10^{-7}}$$

for $\lambda = 6 \cdot 33 \times 10^{-7}$ m for the red line of the helium-neon laser, hence $p = 5 \cdot 26$.

As p must be an integer the highest order obtainable for this light with this grating is 5. The fifth-order beam is diffracted through an angle of about 140°, which gives considerable credence to the concept of Huyghens' secondary wavelets.

All these diffraction grating experiments may be very effectively displayed by using the smoke box referred to on page 3; the grating is mounted on a stand either inside the box or close to its end. Figure 1.39 is typical of the three-dimensional display that is made possible by such an arrangement. It shows that the beams, after diffraction, are emerging parallel-sided from the grating.

If a grating with larger spacing (fewer lines per metre, say $1 \cdot 5 \times 10^5$ lines per metre) is placed in the beam, more orders will be obtained.

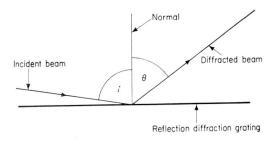

Fig. 1.41. Reflection grating.

It is possible to make rough measurements inside the smoke box, with only a metre rule, instead of the vernier angle measurements made in the traditional diffraction grating experiments using a spectrometer.

A simple reflection grating may be obtained by using a good quality metal rule with the light at more or less grazing incidence (fig. 1.42). The incident beam makes a small glancing angle with the rule such that the beam is spread to cover about 120 of the rule's half-millimetre divisions. The diffracted beams fall on a screen at about 3 m from the rule.

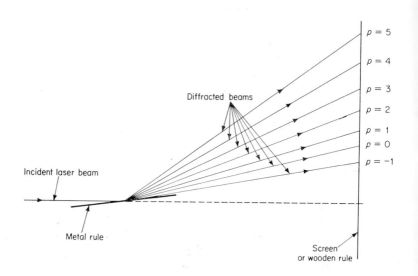

Fig. 1.42. Simple reflection grating.

The equation relating the wavelength λ of the laser light to the angles of incidence i and diffraction θ is $p\lambda = d(\sin i - \sin \theta)$, d is the rule spacing, and p is the order of the spectrum. When $\theta = i$ the order of the reinforcement beam is called the zero order. This is because when $\theta = i$ (fig. 1.41) there is zero path difference between the light beams from neighbouring markings. The various beams are due to diffraction at the rulings on the rule and not to specular reflection. When θ is smaller than i, p is counted positive; and when θ is larger than i, p is negative. The negative orders are usually quite visible unless i is extremely small, and it is worth while to rotate the rule gradually and observe their appearance and disappearance as the angle i is altered. If i is increased, the number of negative orders decreases. The spot on the screen due to the zero order beam is easily identifiable, since it is by far the most intense, and it satisfies the equal angle condition, namely, $\theta = i$.

It is remarkable that two simple rulers, each having 0.5 mm as its smallest division, may be used as the sole measuring devices in an experiment to measure the wavelength of light whose value is 633 nm and to obtain results to an accuracy of about 2 per cent.

1.13. *Fresnel zones and the zone plate*

The zone treatment of Fresnel reconciles the Huyghens' principle with what actually happens when a large wavefront is being propagated. If W (fig. 1.43) is the instantaneous position of a plane wavefront, and the ultimate amplitude at some point P due to W is required, then the different distance from parts of W to P have to be allowed for. The secondary wavelets arising at W will not be in phase with one another at P. However, the wavefront may be divided into zones, with boundaries at distances $(b+\lambda/2)$, $(b+\lambda)$, $(b+3\lambda/2)$, ... from P, each boundary has a phase contribution at P which differs by π from the phase contribution of its neighbouring boundary. The resulting disturbance due to wavelets from alternate zones are in phase with one another. The full calculation, which is not pursued here, shows that with a large wavefront the effects from all but the central half area of the first zone cancel.

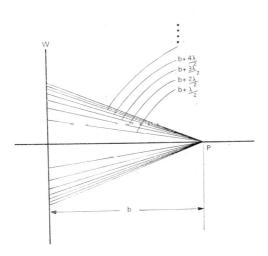

Fig. 1.43. Diagram of Fresnel zones on wavefront W for a point P.

If a set of zones is constructed for P (fig. 1.44 *a*), and the alternate zones blacked out, the wavelets from all surviving zones reinforce one another at P. This device is called a *zone plate* and it is a beautiful application of Fresnel's zone treatment.

A zone plate (fig. 1.45) is a set of circular concentric areas with alternate areas opaque, the remaining areas being as transparent as possible. The radius of the outer boundary of the nth zone is $\sqrt{(nb\lambda)}$,

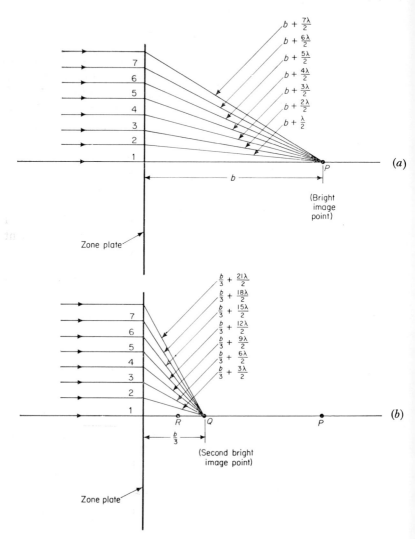

Fig. 1.44. Principle of zone plate.

Fig. 1.45. Zone plate, much enlarged.

so that to make a zone plate circles of radii $a=\sqrt{(b\lambda)}, \sqrt{(2b\lambda)}, \sqrt{(3b\lambda)}\ldots$ are drawn, the value of a determining the value of p; alternate zones are then blacked out, and the whole photographed (as described on p. 5), the reduction process scaling down a and hence p. At the appropriate point P, alternate zones contribute to the full. Other bright points will be found nearer to the plate and on its axis. The next bright point occurs at a point Q (fig. 1.44 b) when each transparent area of the zone plate coincides with three Fresnel zones for this new point Q.

The contributions from each zone may be represented by a phasor diagram*. The direction of the phasor represents the phase of the contribution and the length of the line represents the amplitude. The total contributions from alternate zones are out of phase with one another (fig. 1.46 a) because of path difference. It may be shown that the amplitude contributions from each zone decrease slightly with the distance of the zone from the centre. Therefore the lengths of the lines in fig. 1.46 (a) decrease. For P the contribution from each transparent area is in phase with that due to the neighbouring area. Therefore the sum of the amplitude contributions for P is given by fig. 1.46 (b). The situation for point Q is represented by fig. 1.46 (c). The theoretical zones 1, 2 and 3 are transmitted by the first bright area of the zone plate. Zones 4, 5 and 6 are blocked off by the first opaque area. Zones 7, 8 and 9 are then transmitted by the second transparent area and zones 10, 11 and 12 are blocked by the second opaque area. This process continues as far as there are zones on the plate. The sum of contributions for the transmitted zones contains twice as many zones (1, 3, 7, 9, ...) whose resultant is represented by a phasor pointing to the right as zones (2, 8, ...) phasors pointing to the left. The total effect is an amplitude that is greater than the amplitude for the first zone alone (which is itself twice the amplitude to be expected when no zone plate is present!) Thus Q is a bright 'image point'. Similarly there is

* A kind of vector diagram in which angles represent *phase angles* instead of actual directions in space.

another position R for which five zones coincide with one transparent area on the plate.

In practical terms, when a screen is placed beyond the zone plate, and moved along the axis of the laser beam, a series of points will be found where the laser beam is sharply ' focused '. The brightest of these will be found farthest from the zone plate at a distance f, known as the equivalent focal length of the zone plate. It is called this because if the experiment is repeated for spherical instead of plane wavefronts, then the centre of curvature of the wavefronts, which is the object point, and the bright image point are like conjugate foci for a lens and related by an equation analogous to the lens formula $1/u+1/v=1/f$; here $f=$(*radius of first zone*)$^2/\lambda$. The other points of maximum brightness will be found at distances from the zone plate in the ratios $f/3 : f/5 : f/7$. Normally about four foci will be observed, and better with a plate for which the central zone is opaque. This is because the theory only accounts for summations on the axis, and away from the axis the summation becomes more complex. In particular for a transparent first zone the background illumination close to the axis and hence close to the foci is higher than when the central zone is opaque. Therefore the foci will be less clear for a zone plate that has its central zone clear than for a zone plate that has its central zone opaque.

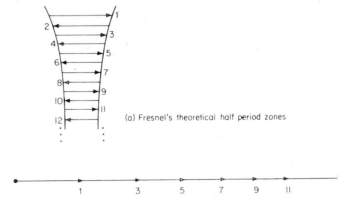

(a) Fresnel's theoretical half period zones

(b) Resultant when alternate zones are blocked by zone plate (point P)

(c) Resultant when sets of three zones are blocked by zone plates (point Q)

Fig. 1.46. Contributions of successive zones.

CHAPTER 2
holography

2.1. *Principles of holography*

HOLOGRAPHY is a method for recording and making available three-dimensional images, using a source of coherent light. The plate on which the necessary information is stored is called a *hologram*.

An object acts as a source or as a scatterer of light waves, which spread out in all directions from all parts of the object. Consider fig. 2.1, which shows two points on an object. An eye at X receives the light from both points and forms an image on its retina. The shape of this image is determined by the relative positions and magnitudes of the two wavefronts as they enter the eye. An eye placed at Y forms an image in the same way, but the relative positions and magnitudes of the two wavefronts entering the eye are different from those of case X, so the two images are also different. The two parameters (variables) that control the relative positions and magnitudes of these waves are their relative amplitudes and phases at the different viewing points.

A hologram records information about both the intensity and the relative phase of the light coming from the object, and if the hologram

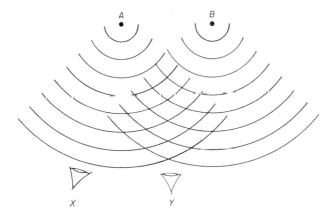

Fig. 2.1. Wavefront pattern from two objects A and B.

is illuminated in a particular way the illuminating beam may be modified with this information to reproduce wavefronts similar to the original. Suppose that an image of a point A is required. Consider the effect on a photographic plate some distance from A. The wavefronts (signal

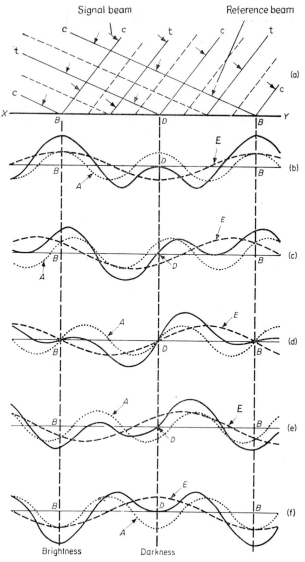

Fig. 2.2. Superposition of two plane waves. The dotted line is the signal beam, the dashed line the reference beam and the full line the resultant summation.

beam) from A are very nearly plane because of the distance from A. Also consider another set of waves of the same wavelength incident on the plate at an angle to the first set. This second set is called the reference beam, and is derived from the same original point in the laser light source as the first set, but its passage to the plate does not involve passing through, or being reflected by, the object. Since the two beams started from the same point, they are coherent and they bear a constant phase relation to each other (see Chapter 6), so that their displacements can be added together according to the Principle of Superposition. Figure 2.2 (a) is a plan view of the two sets of waves arriving at XY, one from the left and the other from the right. A sequence of the relative positions of the signal and reference beams for time intervals of one-eighth of a cycle is shown in figs. 2.2 (b), (c), (d), (e) and (f). The time sequence is for the profiles of the waves at the plane XY. In the plan, the straight lines represent the crests (c) and troughs (t) of the waves, and the dashed lines represent zero displacements. In the profiles, A (dotted) corresponds to the wave from the left on the plan. The wave that produces A is from a point on the object to be holographed, the signal beam. The wave that produces the profile E (dashed) comes from the laser by a path that does not involve the object, the reference beam. The continuous line represents the summation of the waves A and E. At the point B this summation varies during one cycle from a maximum to zero to a minimum and back again. Therefore, B is a position of brightness. At the point D the summation is always zero, therefore D is a dark point. Thus, across the photographic plate, there is a series of light and dark places which (if we extend the argument of the diagram to three dimensions instead of just two) would appear as a series of bright and dark lines at right angles to the plane of fig. 2.2 (a).

Let us apply these ideas to the two-point object and one common plane reference wave. If one point only were present with only the reference wave it would set up a set of bright and dark lines. Likewise for the second, but its interference pattern would be different because of the different directions of the waves. When both are present with the reference wave there are two sets of overlapping but different bright and dark lines, which give a more complex interference pattern. Thus, every point on a complex object sets up an interference pattern with the reference wave. A photographic plate placed in the plane of this complex pattern records the total interference pattern due to all the object waves.

Reverting to the pattern due to a single point again, the spacing of the lines is a measure of the relative change of phase between the signal beam and the reference beam, which is really a measure of the relative directions of the two beams of this example.

This is illustrated in figs. 2.3 (*a*) and (*b*), where the letter D indicates darkness and the letter B brightness. There is a large angle between the beams in 2.3 (*a*), and a small angle in 2.3 (*b*). In (*a*) the B-points are close together, and in (*b*) they are further apart. In this way,

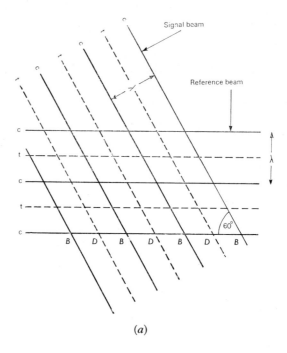

(*a*)

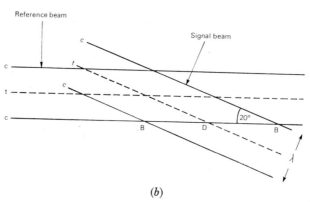

(*b*)

Fig. 2.3. Superposition of reference beam and signal beam.

the spacing of the 'interference' lines is a measure of the relative directions of the two beams.

The summations of figs. 2.2 are for reference beam and signal beam of equal amplitudes. The summations for equal and unequal amplitudes are considered next in two examples for a single point object. In one case, it emits light of amplitude equal to that of the reference beam, and in the other case it emits light of one-quarter of the amplitude of the reference beam. The summations equivalent to those of fig. 2.2 (b) are shown for each of these cases in figs. 2.4 (a) and (b) respectively. The wave notation is as for fig. 2.2; the dashed wave is the reference beam the dotted wave is the signal beam and the continuous wave is the

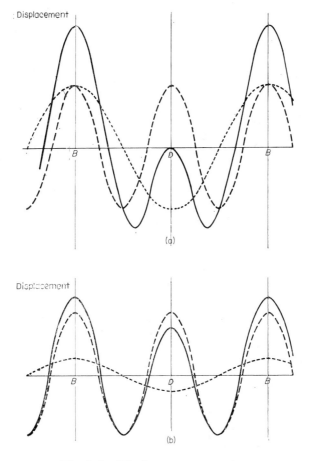

Fig. 2.4. Displacement summation.

39

resultant displacement. These figures show that changes of relative amplitude, (and hence intensity) affect the size of the peaks but not their separation, and change of amplitude affects the contrast but not the spacing of the pattern.

Thus, an interference pattern formed by the combination of signal and reference beams (fig. 2.5) stores information relative to the reference beam about the phase of the light from the object by virtue of the spacing of the fringes and information about the amplitude of the light from the object by virtue of the contrast of the fringes. The information necessary to define the wavefronts emitted by the original object has been stored in the hologram. But this cannot be decoded at sight. Figure 2.5 gives no direct indication of the original object. The next stage in holography is the method of releasing this information, so as to obtain an image from this record.

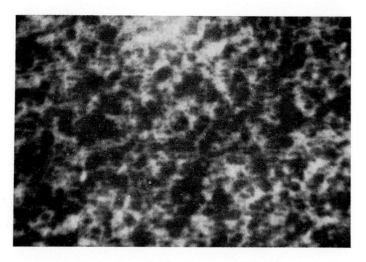

Fig. 2.5. The interference pattern in which information is stored. (This is part of a hologram magnified 500 times).

This may be explained by first considering a diffraction grating illuminated normally by a plane wave that is equivalent to the reference wave of the previous stage. Such a plane wave will proceed from the grating as a plane wave in a definite direction given by $d \sin \theta = p\lambda$, where d is the grating spacing and λ is the wavelength of the 'reference wave equivalent'. There will be various values of θ corresponding to various integral values of p. Another grating in the same relative plane to the reference wave equivalent as the first grating, but whose spacing is $d/2$, produces different values of θ; $\sin \theta$ for a given value of p is proportional to $1/d$. If the same 'reference beam equivalent' illuminates these two gratings side by side then beams of corresponding values of p will appear, one in direction say θ_1 and the other in direction say θ_2. If these two gratings are replaced by a hologram which is really a complex set of millions of diffraction gratings, then millions of emergent beams will appear. Compare this last process with that of making the hologram pattern. The greater the angle X between the reference and signal beams, the closer the "lines" formed on the holographic plate. The larger the number of "lines" per unit length of the plate in the viewing stage, the greater the angle θ between the 'reference beam equivalent' and the emergent beam. It may be proved that this inter-relation is quite precise and that $X = \theta$ if $p = 1$. Thus the various spacings define the emergent directions of the viewing beam, and these directions are the same as the directions of the original waves. The original signal beam that produced the spacing with the reference wave was coming from behind the grating equivalent. The emergent beam as it was originally, therefore appears to come from *behind* this complex grating. Holograms produce virtual images if illuminated in the same manner as when they were made.

The strength of the emergent waves will depend on the relative contrast of the lines of the 'diffraction grating equivalent' that produces the emergent beams. For high contrast, the emergent waves will be strong and for low contrast, the emergent waves will be weak.

The stored information about the amplitude and phase of the original signal beam is converted into emerging wavefronts, whose directions and relative intensities are the same as for the original waves. This information is released by illuminating the hologram with an equivalent reference wave.

2.2. *Three-dimensional images from a hologram*

The eye has no means of distinguishing at all whether the waves it receives come from a hologram or from the original object, hence the image is, like the object, seen in three-dimensional perspective. If the observer changes position, then the relative positions and magnitudes of the wavefronts entering the observer's eye alter and the image changes with the viewing position just as the original object would. This

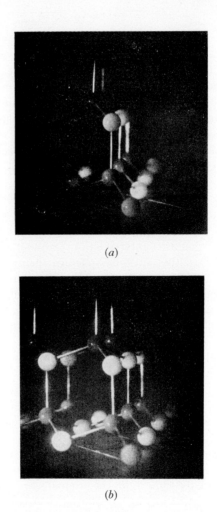

Fig. 2.6 (a) and (b). Demonstrating the three-dimensional effect.

effect is shown in figs. 2.6 (a) and (b), which are photographs taken of the virtual image produced by one hologram. Each photograph is for the camera position marked by its appearance letter in fig. 2.7. Notice that as the camera goes from position (a) to position (b) the object point A appears from behind the object point B. (The photographs cannot really do justice to the effect as seen directly.)

This three-dimensional effect may be shown in another way. In figs. 2.8 (a), (b) and (c), the focus control of the camera has been altered

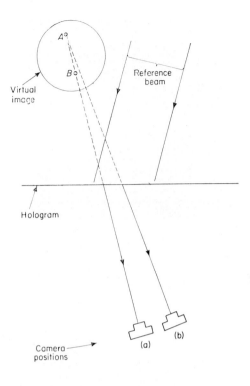

Fig. 2.7.

to pick out in turn the foreground in (*a*), the middle ground in (*b*), and the background in (*c*).

To emphasize the point, the same experiment has been repeated with the same camera being used to photograph a conventional photograph, instead of the image from a hologram. Figures 2.9 (*a*) and (*b*) are for the different camera positions on similar lines to figs. 2.6 (*a*) and (*b*) taken of the hologram image. Figures 2.9 (*a*) and (*b*) show no movement of point A from behind point B. Figures 2.10 (*a*), (*b*) and (*c*) are for the change of focus control, similar to figs. 2.8 (*a*), (*b*) and (*c*), but this time (*a*) and (*c*) are entirely out of focus, and (*b*) is entirely in focus. The essential difference between a hologram and a photograph is that a hologram when properly illuminated recreates the *amplitude* and *phase* relationships of the wavefronts from the original object, whereas a photograph when illuminated only recreates the *intensity* relationship in one plane.

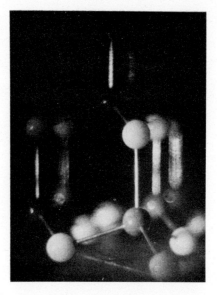

(*a*) Foreground sharply focused

(*b*) Middle region sharply focused   (*c*) Background sharply focused

Fig. 2.8. Different focusing demonstrates the three-dimensional effect.

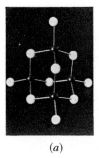 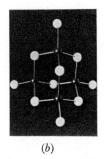

(a)   (b)

Fig. 2.9. A *plain photograph* looks the same seen from two different angles (contrast fig. 2.6).

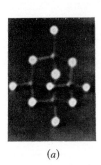 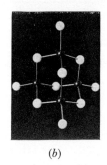 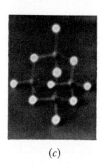

(a)   (b)   (c)

Fig. 2.10. Change of focusing on a plain photograph; the *whole* of (a) and of (c) is equally out of focus (contrast fig. 2.8).

2.3. Transmission holograms

Holograms may be obtained in several ways. These differ largely with respect to the method of illuminating the object, and the subsequent signal beam obtained from this illumination. These different classes of holography are transmission, self and reflection, of which transmission is by far the easiest experimentally, especially with low power (0·25 mW) lasers.

Figure 2.11 outlines an arrangement for obtaining transmission holograms. The signal beam, from the object being holographed, must bear a constant phase relation to the reference beam throughout the exposure of the photographic plate, which may last from 10^{-2} s to several minutes. The output from the laser is spread out to illuminate the whole of the object using a short-focus converging lens, which gives a beam of light diverging from the principal focus F at an angle which depends on the width of the original laser beam, and the focal length of the lens; the shorter the focal length of the lens, the greater the divergence. Microscope objectives provide convenient good quality short-focus lenses. It is desirable to keep the whole of the optical system as

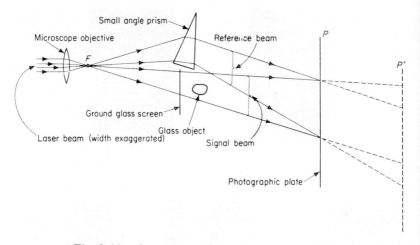

Fig. 2.11. Arrangement for transmission holography.

stable as possible for the duration of the exposure, and this is easier for compact systems, hence the short-focus lens.

This beam is divided into two parts; one half passes through a small-angle (approx. 15°) glass prism. The other half falls on a ground-glass screen, and after passing through the screen illuminates the object. This should be transparent, and suitable examples are a torch bulb or a glass animal figure. The light passes through the object undergoing normal refraction; if it were a plain sheet of clear glass, there would be no pattern to observe, let alone holograph. However, with a shaped glass object the varying amounts of refraction and scatter make it distinguishable. Each point on the object sets up new spherical wavefronts which are a function of the shape of the object. The ground-glass screen provides a background against which the object stands out reasonably clearly. If this were not done, amongst other things, in the background, there would be a bright image point which would be the image of the principal focus F of the lens. This would be extremely dangerous (see the chapter on safety). The grinding of the screen must be very light, otherwise the laser beam will be so much scattered that insufficient light will reach the photographic plate as a signal beam. Since this scatter is only limited the object beam is still largely directional as shown in fig. 2.11. It is desirable to keep the exposure times as short as possible, so that the system has a better chance of being stable for the duration of the exposure. To achieve this with low power (0·25mW) lasers every effort should be made to use as much of the laser output intensity as possible.

The photographic plate P is placed where the overlap of the reference and signal beams is greatest, that is, position P not P' (fig. 2.11). This position is largely determined by the apex angle of the prism used, so by using different prisms, different fields can be observed. The components must be arranged overall so that the path difference between the object beam and the reference beam is less than the coherence length (pp. 96 and 97) of the laser. Ideally, it is best to make the average path for the signal beam equal to the average path for the reference beam. The difference in these paths must not be greater than the coherence length of the laser, because if it is, the phase relation between the two beams is lost. For the low-power cheaper lasers, this length is approximately 40 cm.

2.4. *Photographic plates for holography*

Special photographic plates are used for holography. High definition films are specified by the number of lines per millimetre that are resolvable by the grain of the film. In the early days of holography, the most suitable film available could only resolve the interference lines due to signal and reference beams at an angle of 5° to each other. This limiting value of the angle between the two beams is known as the *design angle* of the system.

The development of holography, with the wish to use large design angles and to work well away from the resolution limit, brought the need for special high-resolution plates. These were first prepared by Kodak, U.S.A., but now most film manufacturers produce holographic plates. These are all capable of being used with design angles close to 90°. Because of their high definition, they must have a very fine grain, and therefore low speed; so the necessary exposure time is long, of the order of seconds or even minutes. This time varies widely, and depends on the exact nature of the system and laser used; it can only be found by trial and error. The holograms used by way of illustration in this chapter vary within the exposure range of 0·5 s to 5 min and were mostly done with a 5 mW laser. Another difficulty is that most reliable cheap continuous wave lasers are the helium–neon lasers, which have a wavelength of 633 nm, a part of the spectrum to which film normally responds slowly. To obtain worth-while response, the manufacturers have to sensitize the film in a special way.

Emulsions for holography are provided in various sizes and on glass or film mountings. It is well to order carefully, bearing in mind the plate holder to be used, and remembering that cutting glass in the dark is not easy. The size of plate holder is not critical, but with output powers of 0·25 mW, an overlap of the two beams of approximately 2·5 cm diameter is more than sufficient. Plates, since they are more rigid than film, are recommended to the beginner. If the size of the hologram is rather wasteful of plates, it is possible by suitable adjustment of the plate holder, in terms of height and/or rotation end for end, to obtain more than one hologram per plate. This can be quite useful in measuring the variation in a hologram due to a particular parameter, the parameter being varied for each exposure and then the plate developed. To estimate the exposure time for a particular arrangement a set of four exposures on different parts of the plate will produce the equivalent of the test strip of conventional photography. The blackening of the plate for a particular illumination is proportional to the *logarithm* of the exposure, so that the range of test exposures should be in the time sequence 1 : 2 : 4 : 8, and not 1 : 2 : 3 : 4. The plates are processed in the usual way. The details of the nature of the developer and fixer solutions and the appropriate times for each of these stages are normally supplied with the plates. Since the plates are slow, the light-proof quality required of the dark-room is not at all high, but there is no suitable safe light for use with red sensitized plates, and all the development must be done in the dark.

2.5. *Viewing transmission holograms*

The simplest way of viewing is to use the same arrangement as was used for taking the holograms (fig. 2.11), except that the ground-glass screen is covered to block off the signal beam, the object is removed and the exposed hologram replaces the plate holder. The hologram must be placed in the same orientation relative to the illuminating beam (the reference beam when it was being exposed). The more carefully this is done, the easier it is for an inexperienced viewer to find the holographic image, by looking through the exposed hologram, *not* in the direction of the reference beam, but in the direction of the original object. He will then see a virtual image of the object at the position which the object originally occupied.

A few pointers here may be helpful.

1. **Do not attempt this viewing until you are sure of the safety precautions to be taken and that you understand the significance of an intensity of** 10^{-2} W m^{-2} **(i.e.** 10^{-6} W cm^{-2}**) in relation to the apparatus being viewed.** In most cases the arrangement of the apparatus and the position of the eye will give a viewing intensity of less than 10^{-2} W m^{-2}. The further back the eye is from the hologram the safer it is, because of the divergent nature of the illuminating beam.

2. Do not look directly into the illuminating beam, since this will produce a bright spot image of the diverging laser beam. By using the same experimental arrangement as that used for taking the hologram it is possible to erect a screen in place of the small-angle prism (fig. 2.11) and this will stop the reference beam entering the observer's eye.

3. If you are at all uncertain as to safety, use a sodium lamp instead. To do this, concentrate, using condensing lenses, as much light from a sodium lamp as possible on to a pin-hole. This pin-hole replaces F of the converging lens in the laser system, and its divergent beam is used in the same way as the divergence of the laser beam. The finer the pin-hole the better the quality of the holographic image, but unfortunately its brightness will also be reduced. The quality of the sodium lamp images will never be as good as those obtained using the laser as described earlier.

4. Focus the eye beyond the hologram, remembering that it is a virtual image that is being looked for.

5. Do not place the eye inside the least distance of distinct vision measured from the virtual image position.

6. If you are disappointed with the first image obtained, remove the hologram, replace the object and the ground-glass screen and look at the object from the direction of the plate holder. This shows that glass objects in a single colour, especially red, are not visually inspiring. **Remember the safety rules when altering the system, because this is when accidental viewing of the divergent beam is most likely.** Nevertheless, these transmission holograms are the easiest to obtain consistently, and the majority of the techniques mastered in this way are applicable to reflection holography which is more interesting visually.

Once the technique of viewing is established, it is not necessary to be so elaborate, apart from the safety point of view, with the viewing arrangements. Images may be produced by a hand-held hologram placed in a *monochromatic* diverging beam, preferably a laser, but a sodium lamp would do. Move the eye from side to side, not backwards and forwards as most people tend to do when looking for the image for the first time. The viewing system does not have nearly as strict a specification as the exposure system. However, it should be a monochromatic beam diverging from a point and stable relative to the hologram to 0·1 of a wavelength for a time which equals the time to form a single image on the retina of the eye of the observer. This last stability condition is much less exacting than the stability condition for exposure, which is that the optical system must be stable to 0·1 of a wavelength for the duration of the exposure of the hologram. The exposure times differ by several orders of magnitude from the retinal image formation time, and it is why lasers are necessary for the exposure to make a hologram but are not necessary for their viewing.

If a more complex reference beam is to be used during the exposure, then the hologram and that reference beam should bear the same relative physical spacing for viewing as that used for the exposure. When some holograms are viewed, a bright spot appears in the image field. This spot is due to too small a design angle which makes the reference beam and the signal beam overlap on the observer's eye retina. This may be overcome by increasing the design angle of the system, or by viewing from further away through a limited aperture entrance pupil.

Figures 2.12 (*a*) and (*b*) are same-size reproductions of two different holograms, (*a*) is a good one, and (*b*) is not so good. Holograms should appear evenly illuminated at this scale of magnification, and to do this the signal beam should be weaker than the reference beam approximately in the ratio of 1 : 10. The signal is only a modulation on the carrier wave, the reference wave. Figure 2.12 (*a*) is not underexposed either; the circled part is the best viewing area. This is because holography is a diffraction process. Holograms are similar to diffraction gratings which generally appear nearly transparent. Figure 2.12 (*b*) has a signal beam which is too strong by about a factor of 2 and shadows appear on the hologram. These are due to the beam-splitting process.

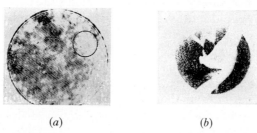

(*a*) (*b*)

Fig. 2.12. Holograms: (*a*) good contrast, particularly in the circled area, (*b*) poor contrast.

2.6. *Stability*

In the past few pages the method of obtaining holograms has been described. Figure 2.13 shows the optical bench used. Much has been said about the severe stability conditions that holography demands. These demands only apply if one is to make precise reproducible measurements, and most holographic work is indeed done at this level of precision. These standards are not necessary for simple experiments which only require that the path differences to the plate should not change by more than one-tenth of a wavelength during the exposure.

This is still a fairly severe requirement, but it may be achieved by firmly clamping all the components to a single metal frame optical bench. Any subsidiary bench movement is then taken care of, at least for exposures up to two minutes. Figure 2.13 shows the laser mounted on normal bench stands, and the microscope objective in a conventional lens holder. An iris diaphragm at the lens focus is used to remove stray light, and to remove any secondary laser beam that may be produced by reflection at the surfaces of the mirrors of the laser. All the beam-splitting parts and the object are placed (perhaps in Plasticine) on a heavy metal plate drilled to take a usual optical fitting. If a metal optical bench is not available then a paving stone or marble slab big enough to accommodate all the optical components will do.

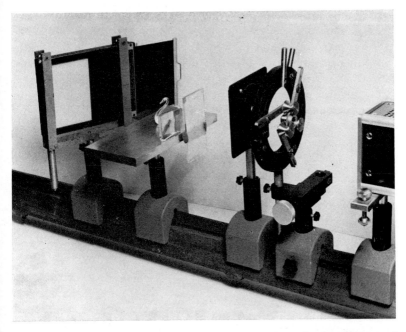

Fig. 2.13. Optical bench arrangement for holography. In this photograph the light comes from the laser on the right.

2.7. Reflection holography

Reflection holography is done on similar lines, and with similar apparatus to that used for transmission holography. Figure 2.14 is a suitable arrangement of apparatus for this purpose. The object is preferably white or red, so that there is maximum re-radiation of the red laser light. Small crystal-model structures are less photogenic but

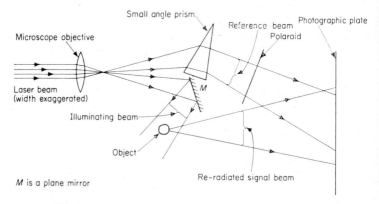

Fig. 2.14. Arrangement for reflection holography.

offer much more interesting detail. The reference beam is obtained as before, except for the introduction of a small square of Polaroid for controlling its intensity. (Further details, pp. 70, 71). The ground-glass screen is replaced by a front silvered plane mirror that reflects the beam on to the object. This object, in turn, re-radiates the light in all directions, some of which arrives at the photographic plate as a complex mixture of spherical wavefronts. Because of the nature of the re-radiation at the object, the intensity of the light in the direction of the photographic plate (signal beam) is considerably reduced in comparison with transmission holography. The reference beam has to have a similar reduction in intensity to preserve the balance of the two beams, and this is the function of the Polaroid which will only work for a laser with a plane polarized output. Otherwise more care must be taken to split the beam in the proportion 1 : 10 as mentioned (p. 50). The result of these reductions in the intensities means an increase in the exposure times of at least a factor of 10, which is why reflection holograms are more difficult to produce than transmission holograms when a low power laser is used.

Other possible arrangements of apparatus use a glass beam splitter, the transmitted beam providing one path, and the reflected beam the other path. A microscope slide placed as in fig. 2.15 is such a possibility. The transmitted beam is over 80 per cent of the incident light, and is best used to illuminate the object. The reflected beam is usually less intense than the incident light, and is used as the reference beam. This is a fairly simple set-up, but there is a danger of multiple images due to multiple reflections at the beam splitter. It is helpful if the total path for the signal beam is the same distance as the total path for the reference beam. The difference in these paths must not be greater than

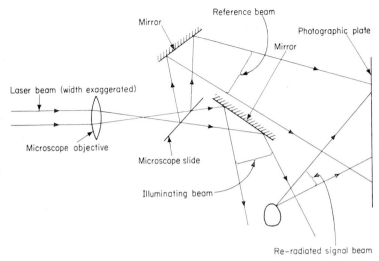

Fig. 2.15. Use of a beam splitter.

the coherence length of the laser, because if it is, the phase relation between the two beams is lost. For the low-power cheaper lasers, this length is approximately 40 cm.

2.8. *Self-holography*

If the ground-glass screen and the prism are removed from the transmission arrangement of apparatus, fig. 2.11, then holograms may still be obtained. In this case the parts of the beam that are uninterrupted by the object act as the reference beam, and the object provides a signal in the usual way. It is because of this possibility of self-holography that the ground-glass screen is used in transmission holography. Self-holography is often the explanation of unexpected images in holograms obtained by other methods. In such cases an unplanned reference beam overlaps the signal beam and produces this extra image. Since this image is unplanned, the chances are that the beam ratios are not ideal. This means that these additional images will be of reduced quality.

2.9. *Multiple image holography*

It is possible that the greatest further application of holography may be in the field of information storage and pattern recognition. The amount of information that can be stored per unit area in a hologram is much greater than that of any other form of storage, even magnetic tape, by several orders of magnitude.

An experiment that shows this feature is to form several images on the same hologram, preferably using the transmission arrangement. The object is a simple two-dimensional one this time, for instance a letter of the alphabet stuck to the ground-glass screen. An exposure is made erring if at all on the side of under-exposure. A different letter is then placed on the screen, the photographic plate rotated through an angle between 5° and 10°, and another exposure is made. This is repeated until the plate cannot be rotated any more. Finally, the plate is developed in the normal way.

When this developed hologram is rotated past a fixed viewing point, each letter appears in turn as the plate is rotated. Up to 200 images have been recorded in this way; with the simple arrangement described here something like five images can be obtained.

Another interesting experiment is to break a hologram, and examine each of the individual pieces. Each piece will be found to produce a *complete* holographic image. This is because each part of a hologram contains a spherical wave contribution from each point on the object. Therefore, information about each part of the object is recorded on every part of the hologram, and consequently each broken piece of hologram shows the full image of the object.

This experiment may be carried out in a less drastic way by placing a variable iris diaphragm (close to the hologram) in the path of the illuminating beam. During the viewing stage, by altering the position and/or size of the iris, different parts and different amounts of the hologram may be viewed. The same effects will be observed as if the hologram were broken into many pieces. The larger hologram has the advantage over the smaller one in that more light is present for a particular image point, and thus the image will be brighter. If very small pieces are viewed, there is a loss of image quality as well as intensity. This brings with it a loss of depth of focus, and is because at the extreme image positions little light is being diffracted, so there is just not enough light to define them.

Figures 2.12 (*a*) and (*b*) have an over-riding diffraction pattern of specks, spots and rings. These are due to particles of dust on one or more of the optical components in the apparatus. There is also a variation in intensity across the exposure. This variation is more apparent on the hologram plate than when viewing a screen placed at the point of overlap instead of the photographic plate. This is because the eye is much more tolerant of intensity variations than is a photographic plate. (These imperfections make the depth-of-focus experiment difficult to do.)

CHAPTER 3
geometrical optics

3.1. Lens aberrations

WHEN the lens formulae are derived geometrically, an important approximation is that $\tan i \approx \sin i \approx i$ and $\tan r \approx \sin r \approx r$, for small angles. If this approximation is not satisfied, then the image of a point object is no longer a point but an area, even if diffraction effects are ignored for the moment.

It may be proved that if the more accurate approximation $\sin i \approx i - i^3/6$ is used, then the image may be represented by five different patterns. For a given pair of object and image positions the image pattern that is observed will be the sum of these five patterns, whose relative proportions are dependent on the shapes of the lenses used, the positions of the stops, the spacing of the various interfaces, and the refractive indices chosen for the various components in the lens system under discussion. These five basic patterns, known as the third-order monochromatic aberrations for a lens or lens system, are Spherical Aberration, Coma, Astigmatism, Field Curvature and Distortion. All five aberrations are normally present in lens systems; however, some of them, or even one, may be predominant over the others.

It is possible to choose particular lenses and use them in particular ways to exhibit a predominant aberration. A laser source may be used for this purpose in two ways: (a) by making an expanded parallel laser beam illuminate the whole of the lens, and hence obtain the total image pattern and (b) by following the narrow laser beam as an individual ray-trace through the lens. The second method needs great care from the safety point of view (see the Appendix), since the laser beam is not spread out at all. This spread is the first safety precaution for many experiments.

3.2. Spherical aberration

To demonstrate spherical aberration, the beam-expanding arrangement described on page 8 gives a suitable broad beam for method (a). Such a beam is incident on a short focal length, large diameter, condensing lens, (fig. 3.1). The emergent beam is not focused to a single

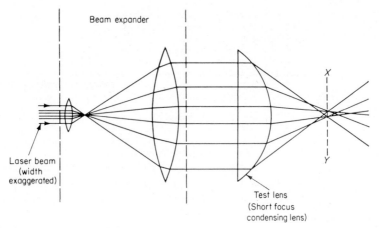

Fig. 3.1.

point, and this lack of point focus for parallel beam parallel to the axis is called spherical aberration (fig. 3.2). The beam area at XY is called the circle of least confusion, since here the beam has its smallest area in the region of focus. Diffraction effects are assumed to be negligible, which is justifiable if the lens aperture is large.

With a test lens that has been more or less corrected for spherical aberration the quality of the image of a point object may be so good that we have to magnify it in order to be able to detect the aberration (fig. 3.3). The magnification, for safety reasons, must be done by forward projection onto an opaque viewing screen, and fig. 3.4 shows a suitable arrangement first described by D. Dutton†. Lenses T_1 and T_2 are

Fig. 3.2. Spherical aberration.

† D. Dutton *et al.*, American Journal of Physics **32**, 355 (1964).

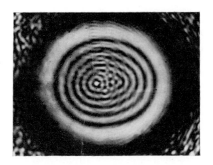

Fig. 3.3. Detail of 'point image' in calculated focal plane, showing aberration.

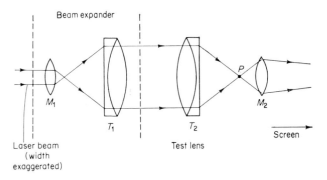

Fig. 3.4.

well corrected telescope objectives (borrowed from a spectrometer for the occasion if necessary). In the diagram these lenses are shown being used in the way their designer, who wanted to reduce spherical aberration, intended; so a nearly 'perfect' image is obtained. M_1 and M_2 are microscope objectives; the shorter the focal length of M_1, the greater is the beam diameter through T_2, which is the lens being examined. Point P is the 'point' image under examination, produced by a parallel beam falling on T_2: M_1 and T_1 provide the expanded parallel incident beam. As mentioned earlier, the details at P cannot be observed with the 'naked' eye, and it would be far too dangerous to examine the regions with a microscope, or even to peer closely. This difficulty is overcome by projecting a magnified image of this image on to a screen, using M_2. Because of its short focal length, M_2 has a limited depth of field, and by movement along the axis the light distribution in the region of P may be examined. When the lenses are arranged as in fig. 3.4, the light patches will vary in diameter down to a point for position P, but there will be no other patterns since the test lens T_2 is being used exactly in the situation for which it was designed to produce little spherical aberration. The light patches will, however, have randomly scattered ring patterns due to dust particles on the lenses, particularly on the microscope objectives. The coherence of the laser makes this more of a problem than it is when the microscope lenses are used in ordinary light. If the lens T_2 is now turned through 180° about a vertical axis, the lens will no longer be used as the lens designer planned and spherical aberration will be observed. Only spherical aberration will be present since the object for the test lens is a point at infinity which is represented to the lens as a parallel beam parallel to the axis of the lens. Figure 3.3 is typical of the images obtained. It is not a perfect picture of spherical aberration; the rings should be circles not ellipses. The elliptical effect is due to slight non-alignment and hence the addition of a small amount of coma (see section 3.3). The lens M_2 is being used to examine the 'point' image a small distance (0·1 mm) away from the best 'point' image.

The approximation $\tan i \approx \sin i \approx i$ decreases in validity the broader the incident beam. This is shown clearly with the help of a variable iris diaphragm, by starting with the iris fully open and then gradually closing it. The image (fig. 3.3) decreases in diameter, and the number of rings also decreases. Eventually the pattern reaches a minimum diameter by which time it has also changed its shape, and it is now the Airy disc diffraction pattern (fig. 1.11, p. 6). If the iris is further narrowed, then this diffraction pattern increases in diameter. This is a particularly important demonstration since it shows that quite large diameter stops, iris size 0·5 cm, produce diffraction effects of the same order as spherical aberration effects for good quality lenses. Diffraction patterns are the realm of physical optics, and spherical aberration is the realm of geometrical optics, yet as this demonstration shows, one fairly coarse adjustment takes the patterns from one category into the other. Thus the relevance of physical optics to the successful interpretation of aberrations is made apparent as far as good quality lenses are concerned.

The extent of the spherical aberration can be measured. A scale is mounted transversely at right angles to the beam. The lens mount moves vertically and transversely. The position of the lens is adjusted until a distant object is focused on the scale, which is then at the Gaussian focus (the focal point for a narrow beam of rays through the centre of the lens parallel to the axis and close to it). The laser is introduced to the optical bench, and the lens is removed from its mounting carefully, so that it can be returned to the original position. The scale reading x cm, fig. 3.5 (*a*) for no lens present is observed. The test lens is then replaced in its mounting, and adjusted until the same scale reading is obtained. This means that the laser beam is passing through the centre of the test lens. The lens is then moved transversely through 0·5 cm, and if the lens were perfect, the scale reading would change by a similar amount to $(x+0·5)$ cm, but this is not so for an imperfect lens. The scale and its mounting are moved along the optical bench until the scale reading is $(x+0·5)$ cm, fig. 3.5 (b). The scale is then at the focus for light 0·5 cm off the axis of the lens. Further off-axis measurements are obtained by traversing the lens in stages across the laser beam, and carrying out axial scale readjustment at each stage. In this way, a set of readings of the change in focal length corresponding to the radius of the aperture of the lens may be obtained.

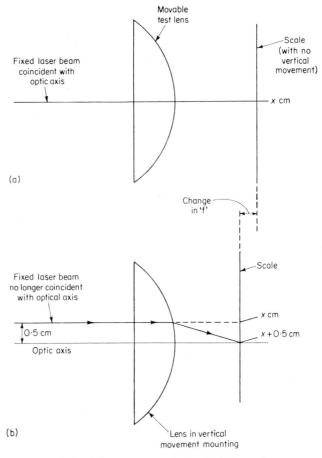

Fig. 3.5. Measurement of spherical aberration.

3.3. Coma

Coma arises for the same geometrical reason as spherical aberration but is concerned with points off the axis. A point object gives an image with a tail, like a comet's, as in fig. 3.6 (a); hence the name coma. Coma with white light gives a shape like fig. 3.6 (a); with a monochromatic source, the tail has a fine structure (fig. 3.6 (b)) due to diffraction effects.

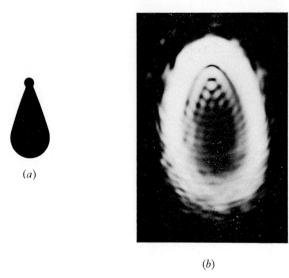

Fig. 3.6. Coma (chiefly).

In order to obtain fig. 3.6 (b), the test lens of fig. 3.4 was tilted fig. 3.7 (a)).

It is possible to show the effect of coma at radial zones of the lens of different diameters by using masks cut to transmit light through one zone at a time. The effect of each single zone is illustrated in figs. 3.7 (a) and (b). Figure 3.7 (a) shows rays from two points in a zone being brought to a point focus; however, other pairs of points in that zone are brought to different point foci. The locus of these foci is a ring, thus for each of the larger zones there is no point image of the point object, but instead a ring image. Zone bb of the lens produces the ring image b. The larger the radius of the zone, the greater is the ring radius, and the further is the centre of the ring from the point image produced by the central part of the lens, fig. 3.7 (b). It is the superposition of these ring patterns that we see as the comet-like image of a bright point in figs. 3.6 (a) and (b).

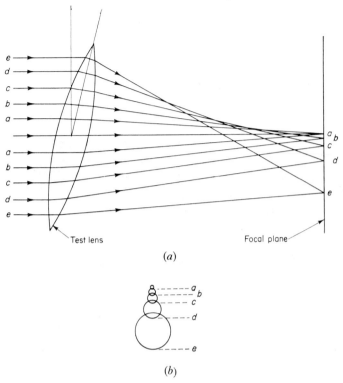

Fig. 3.7. Origin of coma.

Coma may also be observed in the smoke box if a poor quality test lens, such as a condenser lens, is used. The difficulty with this demonstration is that other aberrations are also present; the comet-like image is observable but coma as such is not isolated.

3.4. *Astigmatism*

For larger angles of tilt another aberration, called astigmatism, becomes significant. In the apparatus of fig. 3.4, the coma aberration may be corrected by tilting T_1 and T_2 through about 10° each in the same direction. This cuts out coma, but astigmatism remains. Figures 3.8 (*a*), (*b*) and (*c*) are typical astigmatic images of a point object. The word 'astigmatic' means 'not brought to a point'. These different images are obtained by moving the projection lens M_2 along the axis of the lens system, so that it scans the various images adjacent to the smallest image point. The middle of this range is the image of fig.

3.8 (b), the fine structure again being due to the complex diffraction sum of the various paths through the lens. As the projection lens is brought nearer to the test lens, the pattern fig. 3.8 (b) thins down to the line form of fig. 3.8 (a), and if it moves in the reverse direction the pattern of fig. 3.8 (c) is obtained. The image pattern increases in size and becomes approximately circular after these line positions are passed. These lines are known as the sagittal and tangential focal lines.

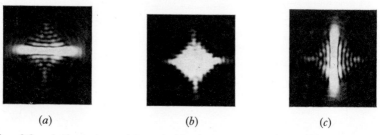

(a) (b) (c)

Fig. 3.8. Astigmatism; (a) and (c) the separated focal lines at different distances from the lens and (b) the intermediate-position astigmatic image of a bright point.

3.5. *Field curvature and distortion*

Spherical aberration, coma and astigmatism are all imperfections of what we should like to be a point image of a point object.

Field curvature and distortion are defects in the image of an extended object, which arise even with many-perfect-point reproductions, since different parts of the object are at different distances from the lens. If a line object is used for a lens which has field curvature on its own its image is sharply focused, but is necessarily curved (fig. 3.9).

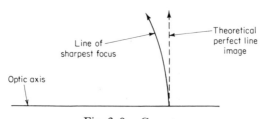

Fig. 3.9. Curvature.

If a square object is used for a lens which has distortion present on its own, then again the image is sharply focused, but this time in a flat plane. However, this image is not the square shape of the original object. There are two types, namely pin-cushion distortion and barrel distortion, figs. 3.10 (a) and (b).

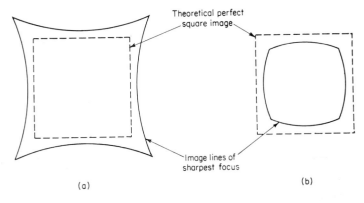

Fig. 3.10. Distortion.

Both field curvature and distortion can be displayed using laser light, but this light has to be spread to illuminate a large object. The directionality of the light has also to be destroyed and this is done by placing a ground-glass screen close to the object, on the laser side of the object. This loses much of the energy of the laser beam, and it spreads what energy is left. The result of this is that the illumination is far less than that present with more conventional methods of examining these aberrations. Thus, while a laser beam is very useful for demonstrating spherical aberration, coma and astigmatism (the 'point' image defects) it is inferior to conventional methods for showing field curvature and distortion.

CHAPTER 4
polarization

4.1. *Introduction*

LIGHT waves are part of the electromagnetic spectrum and their transmission is by electric and magnetic field variation in space. The conventional way of representing an electromagnetic wave is as in fig. 4.1. The directions of the electric and magnetic fields are at right angles to each other and to the direction of propagation of the wave. A wave of the form of fig. 4.1 is said to be plane polarized since the electric field is confined to one plane. The magnetic field is also confined to one plane which is at right angles to that of the electric field. A convention has been adopted that the plane of vibration of a plane polarized wave is said to be the plane that contains the electric field. Imagine looking at the wave of fig. 4.1 along the direction of propagation of the wave, fig. 4.2 (*a*) represents this view of the on-coming plane polarized wave. This may be simplified by ignoring the magnetic field as in figure 4.2 (*b*).

Most light beams, even very fine beams, consist of a very large number of waves of the form of fig. 4.1 with electric fields randomly distributed at various orientations all at right angles to the direction of propagation. This situation is represented in fig. 4.2 (*c*) and the wave is said to be unpolarized.

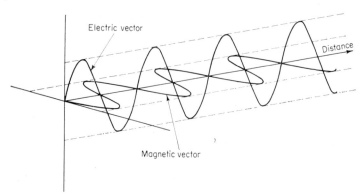

Fig. 4.1.   Representing a plane polarized electromagnetic wave.

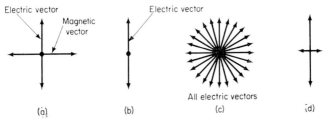

Fig. 4.2. Conventional representations of plane polarized light (*a*) and (*b*); 'ordinary' unpolarized light (*c*); and partly plane polarized light (*d*).

4.2. *Scattering*

A beam of unpolarized light is incident on a small charged particle P. The wave vibrations in the beam set the charged particle vibrating in many directions all at right angles to the direction of propagation PX (fig. 4.3) of the original electromagnetic wave. The vibrations of the charged particle provide a secondary source of electromagnetic waves. Consider the plane YZ (fig. 4.3) at right angles to the original direction of propagation. The vibrations of the particle lie entirely in this plane and therefore the vibrations of the waves travelling in the plane YZ resulting from the vibrations of the secondary source also lie in the

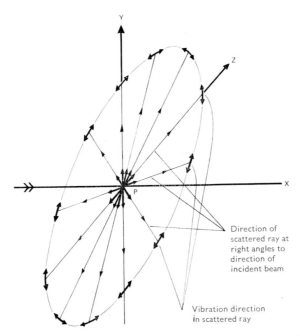

Fig. 4.3. Scattering of unpolarized incident light by small particle at P.

65

plane YZ as shown in fig. 4.3. Thus scattered waves in the plane at right angles to the incident direction of propagation are plane polarized.

This property is used in the following experiment to determine the plane of transmission through Polaroid (privileged direction) of a plane polarized wave. A parallel beam of white light is passed through a smoke box (p. 3). The light scattered by the smoke particles along a direction at right angles to the direction of propagation of the beam is observed through a piece of Polaroid. If this Polaroid is rotated through 360° about an axis parallel to the direction of this scattered light there will be two positions of the Polaroid for which the scattered light will be cut off by the Polaroid. This is because light scattered at 90° to the direction of the initial beam is plane polarized, and the Polaroid only transmits components of the electric field that are parallel to the privileged direction in the sheet of Polaroid. If this privileged direction is at right angles to the plane of vibration of the scattered light, then no light is transmitted by the Polaroid. This provides a method of identifying the privileged direction.

4.3. Brewster windows

It is possible to examine what happens to the plane of vibration of the electric field both when light is refracted and when it is reflected by a piece of dielectric material such as glass. A beam of unpolarized light AB is incident on a sheet of glass as in fig. 4.4 (a). Some of the incident light is reflected in the direction BC. Positions of a piece of Polaroid giving maximum intensity and other positions giving minimum intensity are found by rotating the Polaroid. These maxima and minima (generally not zero) positions are at right angles to each other. The reflected light is partially plane polarized and it may be represented by one long and one short line at right angles, as in fig. 4.2 (d). (The refracted and subsequently emerging light does not show any significant effect with the Polaroid; there is an obvious reason for this.) But there is one particular angle of incidence, the Brewster angle, for which the reflected light is entirely plane polarized. In this situation a Polaroid setting which lets through *no* light can be found. At the Brewster angle, fig. 4.4 (b) it turns out that the angle between the reflected ray and the refracted ray is 90°, which also means that the angles of incidence and refraction are complementary; the refractive index $n = \sin \phi / \sin \phi' = \sin \phi / \cos \phi = \tan \phi$; so ϕ for the Brewster angle is $\tan^{-1} n$. For $n = 1 \cdot 5$, ϕ is approximately 57°. If, having set things to observe the Brewster angle, the direction of the light is reversed so that it strikes the glass after coming through the Polaroid, there will be no reflected light. This means that light plane polarized in this

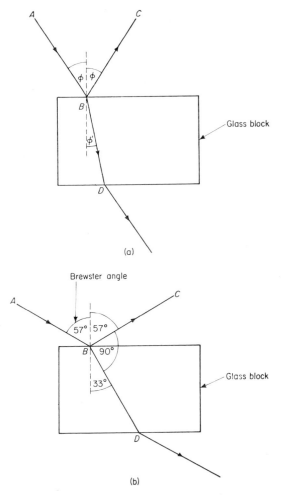

Fig. 4.4. Polarization by reflection.

particular plane incident at the Brewster angle is wholly transmitted by the glass interface. There will be a little absorption in the glass, and some scattering, but these losses are less than 0·1 per cent for a clean surface. A piece of glass set so as to give an angle of incidence 57° for light that is plane polarized in the plane of incidence is very close to being a perfect transmitter. This is a Brewster window. Such windows are often used in gas laser construction, in which case the output of the laser is plane polarized. Some lasers on the market do not use Brewster windows, so that their output is not plane polarized. The state of polarization of a laser output may be readily established with the help of a sheet of Polaroid which is placed in the path of the laser beam its plane being at right angles to the beam, The sheet is then slowly rotated about an axis parallel to the beam; twice in a cycle, at intervals of 180°, the beam will be entirely cut off by the Polaroid. At points in the cycle exactly half-way between the cut-off positions, there will be maximum transmission.

4.4. *The Brewster angle experiment*

The Brewster angle may be measured using a photocell and meter, the reading of which represents the intensity of the incident light and, if the arrangement is linear, is proportional to, the photocell current (I'). It is often assumed that the detector is linear in its current response to the intensity of the radiation. This linearity may be checked as follows. A short focus lens such as a microscope objective is placed in the laser beam path to provide a diverging beam from a point source (fig. 4.5) which is the principal focus of the lens. Values of I' are observed for different distances x (fig. 4.5), and a graph plotted for $(I')^{-1/2}$ against x. If the graph is linear, this is evidence of the linear response of the detector. Since if

$$I' \propto \frac{1}{(x+d)^2},$$

$$(x+d) = c \times \frac{1}{\sqrt{I'}}, \text{ where } c \text{ is a constant};$$

$$\therefore \quad x = \frac{c}{\sqrt{I'}} - d.$$

If the graph were not a straight line, the inverse square law could be used to calibrate the response of the meter in terms of the intensity of the incident radiation.

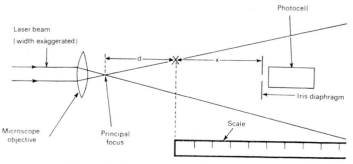

Fig. 4.5. Inverse square law experiment.

The values for d and x should be chosen so that the photocell lies well inside the cone of radiation since the laser beam output decreases in the peripheral sections of the beam.

The beam for the Brewster angle experiment is not divergent, therefore the distance from the reflecting surface to the photocell does not have to be maintained constant. A microscope slide is rotated about an axis through B (fig. 4.6) at right angles to the laser beam until a position C is found which is the position of maximum response, and it will be when the angle of incidence equals the angle of reflection.

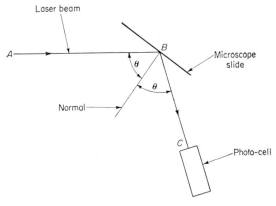

Fig. 4.6.

With the plane of vibration in the laser beam in the plane of incidence, the position of the photocell is moved and the slide rotated again. This is repeated until a position of the photocell is found for zero response throughout the rotation. The Brewster angle will be half the angle between the incident laser beam and the reflected laser beam entering the centre of the photocell aperture in this response situation. This experiment may be shortened by assuming that the Brewster angle for glass is around 57°, and setting the glass slide at this angle. This slide is then rotated about the axis of the beam, until there is no reflected light. The plane of vibration in the beam is now that which includes the laser beam and the normal to the slide. This is likely to be close to the vertical plane, since the Brewster windows inside lasers are often mounted in a plane that is close to 57° to the horizontal. With some lasers it is possible to look down through the top of the laser, through ventilating holes in the roof of the housing, and see how the windows are situated. **(The laser must be switched off while this is being done.)**

4.5. Law of Malus

This experiment can be developed into a more detailed analysis of the intensity of the light transmitted by the Polaroid at any position between the two extremes. The Polaroid is mounted close to a circular scale (fig. 4.7) made from protractors taped to a sheet of glass or Perspex. For class demonstration it is probably best to use the Polaroid sheet in the shape of a square, using the corners as reference marks.

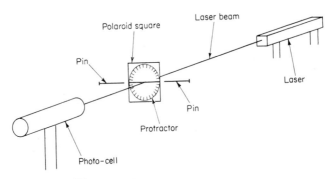

Fig. 4.7. Transmission by Polaroid.

The Polaroid and protractors are placed in a laser beam. Two pins are placed at the sides of the protractors to provide zero points for the scale position. Beyond the Polaroid is a photocell and meter, the reading of which represents the intensity of the light falling on which, if the arrangement is linear, is proportional to it, the photocell current. This may be repeated for different orientations of the Polaroid. The zero lines of the protractors are made parallel to the axis of maximum transmission, and the photocell is used to assist in making this setting. The photocell current I' (assuming linearity) is then related to the angle θ between Polaroid and reference direction by the expression, $I' = I'_0 \cos^2 \theta$, where I_0' is the maximum current. I' may be plotted against $\cos^2\theta$, when the result should be a straight line through the origin. The results of this experiment provide evidence for one of the fundamental wave properties, that is, the intensity of a beam of waves is directly proportional to the square of the amplitude of the waves. It has been assumed here, first that there is a preferential transmission direction for Polaroid, and second that amplitude may be resolved like a vector quantity. In fig. 4.8 MN is a plane polarized wave of amplitude A, vibrating in a plane at an angle θ to the preferential transmission direction of the Polaroid. The component of A parallel to the preferential direction is $A \cos \theta$. If the intensity of the beam is

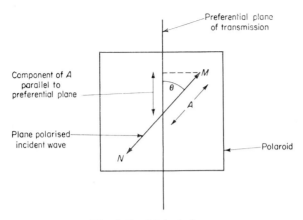

Fig. 4.8. Malus's law.

proportional to the square of the amplitude, then the intensity associated with this component is $kA^2 \cos^2 \theta$, where k is a constant of proportionality. When $\theta = 0$, $I_0 = kA^2$, therefore $I = kA^2 \cos^2 \theta = I_0 \cos^2 \theta$. The resulting straight line through the origin confirms that amplitude is a vector quantity and that the intensity is proportional to the square of the amplitude. The $\cos^2 \theta$ relationship for the intensity transmitted is known as *Malus's law*.

4.6. Double refraction

Some features of double refraction are best shown with unpolarized light. However, the intensity of the laser beam and its plane polarized condition are useful for some experiments on double refraction. If a plane polarized laser beam is passed through a calcite crystal, then usually two beams are seen emerging at the far surface of the crystal, arising from a single incident beam. With some samples of calcite crystal, there is sufficient scatter inside the crystal to see the beam travelling as two separate beams from the surface on which it is incident. They diverge further as they pass through the crystal. By examining different samples of crystal, it may be seen that the separation of these two images is directly proportional to the length of the crystal. When the crystal is rotated in its holder, with the light incident normally to one of two parallel surfaces, then one of the emergent beam positions appears to be stationary (known as the ordinary ray), and the other beam (the extraordinary ray) position rotates about the first position. Each of the beams is plane polarized from the point where they separate at the first surface.

The amplitude of the vibration in each of these beams is equal (apart from losses due to absorption and scatter) to the component of the incident vibration amplitude that is parallel to the plane of vibration of the refracted beam. As the crystal rotates in an anti-clockwise direction, fig. 4.9 (a), (b), (c) and (d), the components of the amplitude of the incident light XY in the directions AB and CD change. In fig. 4.9 (b) these components are $A \cos \theta$ and $A \sin \theta$ respectively. The first component of the amplitude, $A \cos \theta$, goes from a maximum in (a) to zero in (d), then another maximum in the plane at 180° to (a), and another zero at 180° to (d), before the rotation brings the crystal back to (a) again. The other component varies in a similar fashion, but its maximum is at position (d) and zero at (a). If the experiment is done for unpolarized light there is no variation in the intensities.

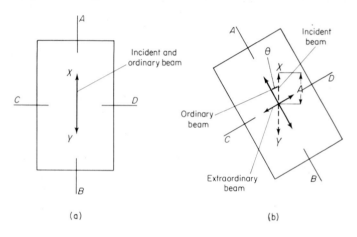

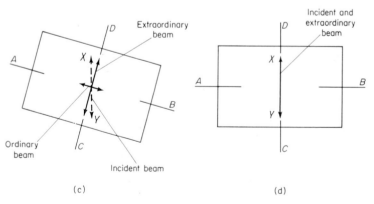

Fig. 4.9. Plane polarized light and calcite.

4.7. Optical activity

Certain solid substances and also a number of liquids have the property of rotating the plane of vibration of the electric vector of plane polarized light ('rotating the plane of polarization'). This is known as optical activity. It may be demonstrated by making a solution of dextrose, concentration of 10^3 kg m^{-3} and using it to fill a long narrow Perspex trough. A laser beam passes down the length of this trough, its path being shown up by natural scatter which is, however, rather directional. A piece of Polaroid is inserted in one end of the trough, and rotated until extinction is observed. More precise observation of the extinction position is obtained if an opaque viewing screen is placed in the tank beyond the Polaroid. The Polaroid is then moved slowly down the length of the trough, and it has to be rotated as it goes to maintain the extinction of the beam. This corresponds to the rotation of the plane of polarization of the beam.

With the protractor arrangement of § 4.5, and a trough of larger cross-section we can measure the angle of rotation θ, the length of the solution l and the concentration m, which are related by the equation

$$\theta = kml,$$

where k is a constant.

It should be noted that θ can be more than 360° and also that some substances produce a rotation in a left-handed direction (laevo-rotatory or anti-clockwise rotation when looking towards the on-coming beam); other substances produce a right-handed rotation (dextro-rotatory or clockwise rotation when seen looking towards the on-coming beam). Suitable liquids to demonstrate the effect are, for left-handed rotation, laevo-fructose, and for right-handed rotation, dextro-glucose. These are both types of sugar, and the optical activities of their solutions decrease with time, a decrease of about 50 per cent in 5 hours.

CHAPTER 5
light emission and energy levels

5.1. *Introduction*

IN the 18th century, Newton imagined that light might consist of corpuscles travelling in straight lines. During the 19th century the wave theory of light, culminating in Maxwell's electromagnetic theory, became established. In 1900, Planck proposed that the energy emitted by a radiating body was not emitted continuously, but in discrete units which he called ' quanta '. This led to the development of the Quantum Theory.

During the early part of the 20th century, de Broglie and others were working on the theory and experimental verification of the idea that electrons have wave properties as well as particle properties. This they achieved, the experimental verification coming from the famous experiments of Davisson and Germer (1927) and G. P. Thomson (1928).

This concept of a duality of properties which was developed for the electron was also used to reconcile conflicting aspects of the behaviour of light. After years of work, argument and discussion, the contradictory theories of the beginning of the century were resolved into one theory in which a quantum of light energy is known as a photon and has both wave and particle properties. This duality of roles

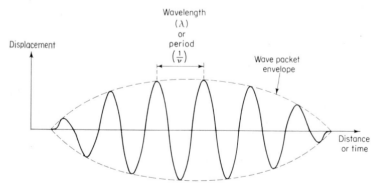

Fig. 5.1. Wave packet representation of a photon.

was explained entirely in mathematical terms, but fig. 5.1 provides a simplified pictorial representation. The frequency (ν) of this wave is related to the energy (E) of one of its quanta by the equation $E = h\nu$, where h is the Planck constant.

The work of Rutherford, Bohr and many others established that an atom has a positively charged heavy nucleus of radius about 10^{-15} m and that its electrons are separate from the nucleus at distances of order 10^{-10} m and specify precise energy levels. The energy of an electron, whilst it remains at one of these levels, remains constant, and no radiation is emitted whilst the electron is at that level. The energy levels for the electrons in a hydrogen atom are shown in fig. 5.2. Levels E_0, E_1, E_2 . . . . . are specified by principal quantum numbers $n = 1, 2, 3$.

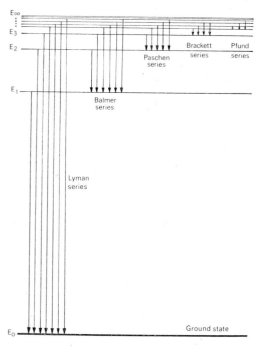

Fig. 5.2. Hydrogen atom; energy levels and early terms of five spectral series.

It was also stipulated that if an electron changed from a higher energy level E_2 to a lower level E_1, then a precise quantity of energy is emitted in the form of a photon as in fig. 5.3. The frequency of the emitted photon is precise and its value may be calculated from the following equation, $E_2 - E_1 = h\nu$, where h is Planck's constant and ν is the frequency. In addition the converse process applies when energy is supplied to an atom. One of its electrons stores this energy by absorbing it and moving from a lower energy level to a higher level. This is illustrated in fig. 5.4. An electron can only absorb the precise quantity of energy which is equal to the difference in energies involved in the change, e.g. $E_2 - E_1 = h\nu$. The supply of energy to the atom does not have to be in the form of a photon of light energy but it is frequently so in the case of laser action.

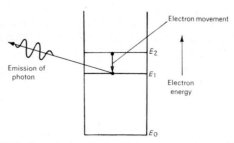

Fig. 5.3. Emission of photon, involving energy levels E_1 and E_2.

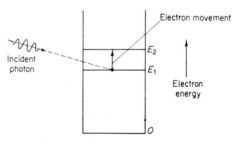

Fig. 5.4. Absorption of photon, involving the same energy levels E_1 and E_2.

The inter-relationship between absorption and emission of photons is illustrated in the following experiment, which uses a spectrometer, diffraction grating or highly dispersive prism, sodium flame and a white light. The spectrometer is arranged to be in normal adjustment and illuminated by white light only. The photons emitted by the white light source have energies that correspond to the whole of the visible part of the spectrum and beyond, that is, a range of frequencies is obtained. Two lenses, fig. 5.5, are used to focus this light on to the slit of the collimator of the spectrometer to give brighter illumination. The grating and telescope are adjusted until the spectrum due to the white light source is clearly visible and it will be of the form shown in

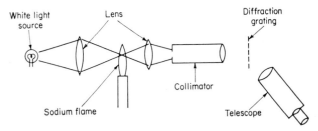

Fig. 5.5. Apparatus for observing reversal of the sodium D lines.

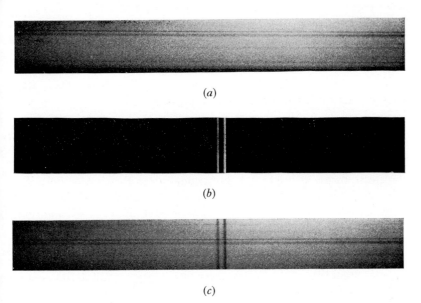

Fig. 5.6. Reversal of the sodium D lines; observation.

fig. 5.6 (*a*), that is, a continuous spectrum with no breaks. The sodium flame is then introduced between white light source and slit at the intermediate image produced by the first lens (fig. 5.5) and the white light is switched off. The spectrum looks like that of fig. 5.6 (*b*), which shows the yellow sodium D lines. If the white light is switched on again, the spectrum will be as in fig. 5.6 (*c*) which shows two breaks in an otherwise continuous spectrum. These breaks are absorption lines, and the wavelengths corresponding to these breaks are found to be exactly the same as the wavelengths of the sodium D lines. If a low dispersive power prism were used instead of the grating, these two breaks would merge into one or might even be completeley obscured through lack of sufficient resolution. This is an interesting experiment, since it gives experimental evidence in support of the quantum ideas for the excitation of atoms. In the case of absorption, the sodium atoms are bombarded by photons whose frequencies range completely across the visible spectrum. The sodium atoms become more excited than if the flame alone were used, and do so by absorbing additional energy from these incident photons. The photographs show that for absorption to take place the incident photons that are absorbed must have the same energy as that required for the atom to be excited, otherwise other lines would disappear from the spectrum when absorption takes place.

The theories described so far in this chapter have only referred to single electrons in an atom and have not mentioned the effect on one another of the other electrons surrounding the nucleus. Pauli proposed his Exclusion Principle which in a simplified form states that there is a specific maximum number of electrons that may be present at a given energy level for a particular atom. These numbers were two for the lowest level, eight for the next and so forth; they provide a theoretical link with the periodic table. If an atom has all its electrons at the lowest possible energy levels as specified by the Pauli Exclusion Principle, then that atom is said to be at the ground state. This does not mean that all the electrons are at level $n=1$, but it does mean that level $n=1$ is full, followed by level $n=2$ being full, and so forth until there are no more electrons left for that atom. If the atom is then supplied with additional energy, for example, by collision with another atom or by heat, then this additional energy must be stored in some way. To do so, some of the atom's electrons may go to higher energy levels. The atom is then said to be excited, that is, the lowest energy levels are not full to the capacity defined by Pauli's Exclusion Principle.

Another step forward was the introduction of a new mechanics which is known as *Wave Mechanics* and is derived from a new set of assumptions. The most basic assumption is that a small particle may be represented by the Schrodinger equation which has the same mathematical form as the equation for a progressive wave (or a stationary wave pattern). The details of this equation are beyond the scope of this book, but one of the most important concepts in the subsequent development of wave mechanics is the Heisenberg Uncertainty Principle. This may be expressed in the form

$$\Delta E \cdot \Delta t \geqslant h,$$

where ΔE and Δt are the uncertainties in simultaneously measuring energy and time respectively, for an electron in an orbit or, as is more often referred to now, a stationary state. Imagine for the moment that it is possible to observe closely an individual electron, and that it is required to measure the energy of the electron at a particular instant in time. In any experiment there is an uncertainty (error) in measuring the quantities concerned. In one's initial experimental experience it is always assumed that a better experiment would reduce these uncertainties. However, the uncertainty principle states that there is a limit beyond which the experiment cannot be improved. In the imaginary experiment described this limit is the product of the uncertainties in measuring the energy and the time. This product cannot be less than h, the Planck constant. It is well to remember that h is an extremely small quantity (6.62×10^{-34} J s) and therefore this principle is usually only of significance when dealing with particles of atomic or sub-atomic size. Also note that if t were measured precisely E would be unknown at that instant, and conversely if E were measured precisely then t would be unknown.

Because of this uncertainty we have to work in terms of a probability; and the precise energy of the quantum theory is now replaced by a probable energy range. The probability distribution is in the form of the Schrödinger wave equation.

Davisson and Germer, in their famous experiment, showed that electrons do have wave properties, thus providing supporting evidence for this new mechanics. The electron microscope is an example in which these wave properties are used directly. It was proposed by de Broglie that the model of an electron rotating about a nucleus in a fixed orbit be replaced by the concept that a particular electron lies somewhere about the nucleus and that its most probable position can be defined by its wave pattern. Schrödinger had previously derived a general wave equation and the mathematical form of this pattern is one of the solutions of this equation. Thus instead of jumps between precise energy levels, it is possible to have jumps between members of a range of energy levels that constitute an energy band. It is important at this stage to keep orders of magnitude in mind. The width of these energy bands in energy units is only a fraction of one per cent of the energy difference between neighbouring bands, which is why the concept of the Bohr atom is acceptable at a simplified level of interpretation.

These energies are specified by states, levels or bands, and are nearly precise quantities. If an electron goes from one energy state to another there is an emission or absorption of radiation of energy equal to the energy difference between the states.

The discussion so far has limited itself to individual atoms. But the quantized energy state concept applies to other situations, for example, the energies of vibration and of rotation of the atoms in diatomic molecules are quantized states and account for the majority of emission lines in the middle and far infra-red.

5.2. *Stimulated emission of radiation*

In normal circumstances an excited atom does not stay excited for very long, and the average time spent in an excited state depends on the energy released in the transition from it. If this amount is large, which corresponds to the X-ray end of the spectrum, the lifetime in the excited state is small; conversely, the lifetime is long for the microwave end of the spectrum.

The excited atom only stays excited for a short time, about 10^{-7} s for the visible region of the spectrum. This time depends on many factors: the temperature, the number of atoms excited at this and other states and the values of the energy jumps involved. Except in certain states this excited lifetime is very short indeed. The excited atom may give up its energy in several ways:

(*a*) by emission of all the energy as it returns directly to the original ground state;

(*b*) by emission in several steps before reaching the ground state;

(*c*) by collision with, and hence the transfer of the energy to, another single atom;

(*d*) by feeding energy to the lattice of a solid where it increases the lattice vibrations, and thus raises the temperature;

(*e*) by *stimulated emission*, which means the premature emission of the photon of frequency v, stimulated by *another* photon of frequency v after a time very much shorter than the ordinary lifetime.

Case (*a*) is known as spontaneous emission. The energy given out as radiation is equal to the energy originally absorbed, therefore the wavelength of the emitted radiation is the same as the wavelength of the absorbed exciting radiation. In the sodium flame experiment, the atoms that absorb photons from the incident white light lose this additional energy fairly rapidly as radiation, but this new radiation is emitted in all directions. There is a net loss of radiation of this particular wavelength from the path of the beam since the portion which was absorbed is now re-radiated in all directions by the excited sodium atoms. The direction, plane of polarization and precise instant of emission are all random.

Flames give one method of producing line spectra. But the excitation energy is often obtained in other ways; for example, in a sodium discharge lamp some of the gas atoms are ionized, and the electric field responsible for the ionization also makes the positive ions move towards the cathode and the electrons move towards the anode. Collisions take place between the electrons (and positive ions) and the unexcited atoms and some kinetic energy is transferred to the atoms and excites them. These excited atoms, after a very short time, return to the unexcited state, releasing the exciting energy as radiation.

Case (b) implies many possible transitions but well-established quantum 'selection rules' predict that in fact most of these are nearly *impossible*, i.e. they are very improbable. These 'forbidden transitions' sometimes occur in spite of the selection rules. Considering the *allowed* cases, each transition in the step-by-step process still involves energy $h\nu'$ where ν' must be less than the exciting frequency, ν, since intermediate transitions involve less energy change than the major one. Sometimes photons are not emitted, the energy being dissipated by processes of types (c) and (d).

Case (e) is, from the point of view of the laser, the most important emission process. It was generally overlooked for forty years after Einstein first proposed the idea. In this case, a second photon of the same frequency is produced when a photon of frequency ν, such that $h\nu = E_2 - E_1$, passes close to an atom at the excited state E_2 and interacts with that atom. The phase, direction, state of polarization and frequency of the second photon all have the same values as those of the incident photon, which is not changed by this interaction; so a second photon joins the first. If the original photon should travel along a tube which is long compared with atomic dimensions, it would interact with many excited atoms and be joined by a large number of photons, all in phase with each other (fig. 5.7).

Typically in a helium–neon (He–Ne) gas laser, the gas pressure is of the order of 1 mmHg, which corresponds approximately to a mean free

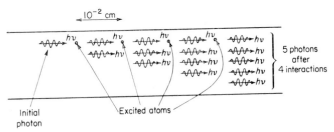

Fig. 5.7. Photon multiplication.

path of 10^{-4} m. If all the atoms are excited to one particular state, then one photon travelling from one end of the laser tube to the other, a distance of the order of 1 m, would have 10^4 interactions. The additional photons would themselves interact with other excited atoms to produce more photons.

In practice the amplification by stimulated emission is not as high as implied here since many of the interactions will be with unexcited atoms. The most valuable factor for lasers in stimulated emission is the production of further photons of the same phase, direction, state of polarization and frequency as the original photon.

5.3. Population inversion

So far it has been assumed that every atom with which a photon interacts is already excited. In a gas in thermal equilibrium it is more likely that a photon will ' collide ' with an unexcited atom and, if so, it will probably absorb the incident photon and become excited. The Maxwell–Boltzmann distribution provides the relation between the numbers of unexcited and excited atoms of a given energy for a particular temperature where the system is in thermal equilibrium. The Maxwell–Boltzmann distribution (originally calculated for the kinetic energy of gas molecules in the kinetic theory of gases) states that for any two energy states E_1, E_2 ($E_2 > E_1$) the numbers N_1, N_2 in those states *under equilibrium conditions* at temperature T is given by $N_2/N_1 = \exp[-(E_2 - E_1)/kT]$.

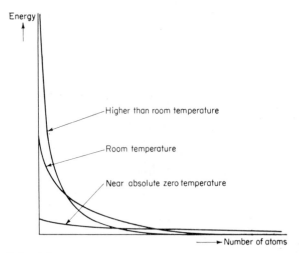

Fig. 5.8. The Boltzmann distribution for different temperatures.

This is called the normal population of the states; note that N_2 is always less than N_1. At absolute zero there should be no excited atoms. As the temperature is increased, the number of excited atoms increases. Figure 5.8 shows the variation with temperature of the distribution of atoms at different excitation energies for a fixed total number of atoms. As the temperature increases the number of excited atoms increases and the number of unexcited atoms decreases. The negative slope of the curve indicates that the distribution never allows the number of atoms at a higher energy to be greater than the number of atoms at a lower energy. Always in a normal population of atoms, it is 'the higher the fewer'. Since energy comes in quanta, the curve is not really smooth as in fig. 5.8, but represents a series of steps as in fig. 5.9, where the dotted line represents the Maxwell–Boltzmann distribution and the rectangles represent the total number of atoms in a particular state, the short side of the rectangle represents the magnified probable range of energies for this state and the long side represents the number of atoms in that state. The most important consequence of the distribution of fig. 5.8 also applies to fig. 5.9. A photon traversing a system in thermal equilibrium is more likely to meet an atom at the lower energy level E_1 than an atom at the higher energy level E_2, therefore absorption is more likely than stimulated emission. To reverse this likelihood, it is necessary to upset the normal distribution artificially (upsetting thermal equilibrium in the process) and so provide more atoms at the higher energy level E_2 than at the lower level E_1. This process is known as *population inversion*, and the distribution then follows the form of

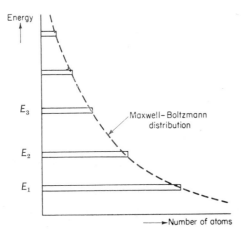

Fig. 5.9. Normal level-population distribution.

fig. 5.10. Stimulated emission is more likely than absorption for the transition from energy E_3 to energy E_2 but not for changes from E_3 to E_1 or E_2 to E_1 since only in the first of these cases is the probability of a photon 'colliding' with an atom greater than the probability of its 'collision' with an unexcited atom.

Absorption and emission between the states E_3 and E_2 both follow the equation $h\nu = E_3 - E_2$; and no other value of $h\nu$ will be concerned in either the upward (absorption) or downward (emission) transition. So when dealing with radiation associated with this transition it is the populations of the two states E_3 and E_2 relative to one another that determines whether stimulated emission or absorption predominates. The populations of the other states play no direct part in this; indirectly they may decide the numbers present in the states E_3 and E_2, but the numbers in these two states alone decide which shall be the main effect. As stimulated emission proceeds, it reduces the proportion of the atoms in state E_3, unless the population inversion is maintained artificially.

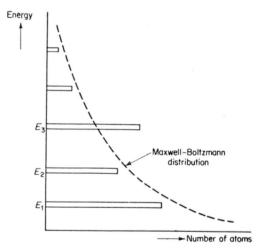

Fig. 5.10. Inversion of level-population distribution.

This is what a laser does and its name is derived from the initial letters of 'Light Amplification by Stimulated Emission of Radiation'.

The numbers of excited atoms in states E_3 and E_2 are usually so close that one traverse through the laser tube does not give sufficient amplification to provide the very great increase in output associated with lasers. To improve on this, each end of the laser tube is made as a highly reflecting mirror, so that the photons are reflected back and forth along the tube, producing more and more photons which are in phase with one another. If we omit the amplification part, this state of affairs is similar to the acoustic stationary wave system in a resonance tube. Usually, in a resonance tube, most of the energy of the stationary wave system is retained in the tube, though some of the energy is allowed to escape. In lasers, one of the end mirrors is not quite as perfect a reflector as the other; for example, one reflects 99·5 per cent of the photons incident on it and transmits 0·2 per cent (the remaining 0·3 per cent is lost in absorption at the mirror surface). Thus a small fraction of the light escapes from the laser tube through this mirror as the laser beam. These figures are typical, but actual values depend on the design used.

CHAPTER 6
coherence

IN this chapter the idea of coherence is explained. The chief beauty of the laser as a source for physical optics experiments is that its radiation is very coherent across the whole width of the beam.

6.1. *The Young's slits experiment*

Thomas Young in 1801 used pin-holes for what is now called the 'double-slits' experiment (fig. 6.1; the dimensions of this figure have been much exaggerated at right angles to direction SS_1). A condensing lens L gathers as much light as possible from a sodium lamp S on to a fine slit S_1. According to Huyghens' principle S_1 can act as a secondary source radiating light through an angle up to 180°. In the plane of the diagram, the wavefronts are arcs, W. The light is diffracted through a large angle, although there is a gradual decrease in the energy per unit area of the diffracted wavefront as the angle θ increases (this discussion applies if the slit S_1 is narrow enough; otherwise the single-slit diffraction pattern is superposed).

The wavefront from S_1 then reaches the slits S_2 and S_3 which are accurately parallel to each other and to slit S_1. The two parts of each

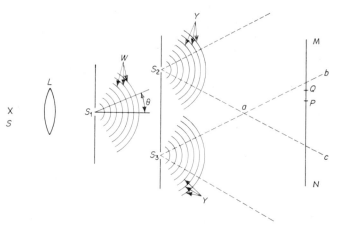

Fig. 6.1. Photons in two-slit interference experiment.

wavefront that reach S_2 and S_3 bear a constant phase relation with each other. S_2 and S_3 act as secondary sources according to Huyghens' principle and the waves from the two overlap in the region *abc*, having approximately equal amplitudes. The resultant displacement at a point at any moment is the sum of the individual displacements at that point. Consider a point P, fig. 6.1, such that the difference between the paths S_2P and S_3P is equal to an exact number of wavelengths. Waves will arrive in step and the resultant amplitude will be twice the amplitude due to one of the waves. This summation of the waves is shown in fig. 6.2, where the two waves are in step because of the amount of the path difference. At another point Q, the path difference, may be an odd number of half-wavelengths, i.e.

$$S_3Q - S_2Q = (n - \tfrac{1}{2})\lambda,$$

where n is an integer, and the waves are now out of step. Then the summation of the waves is zero as shown in fig. 6.3. These two cases are often known as constructive and destructive interference respectively. In the general case which is shown in fig. 6.4, the resultant wave

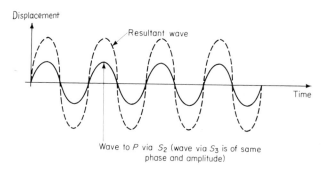

Fig. 6.2.

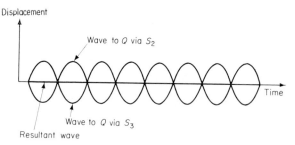

Fig. 6.3.

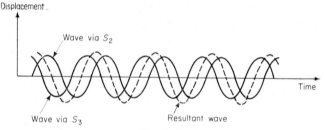
Fig. 6.4

form is still apparent, but the amplitude is less than the maximum which is found when there is exact constructive interference. Each of figs. 6.2 to 6.4 shows a plot of the variation of displacement with time at a fixed point of observation. In fig. 6.2 the resultant displacement varies considerably, that is, a wave passes the point P. No detector is capable of following these variations with time. The largest variation corresponds to maximum brightness and absence of variation corresponds to darkness. If a screen MN is placed in the region of approximately equal amplitude (*abc*) in fig. 6.1 then a pattern of dark and bright lines is obtained (fig. 1.20, p. 11).

6.2. *The average number of photons in an interference pattern*

Since $h = 6 \cdot 62 \times 10^{-34}$ J s and $c = 3 \times 10^8$ m s^{-1}, a single photon of light of wavelength 600 nm has an energy

$$h\nu = hc/\lambda = (6 \cdot 62 \times 10^{-34} \times 3 \times 10^8 / 600 \times 10^{-9}) \text{J} \approx 3 \times 10^{-19} \text{ J}.$$

Let us again consider the Young's slit arrangement of fig. 6.1 and carry out a 'thought experiment'. A lens of diameter 100 mm might be used, the lamp at twice the focal length of the lens away from the lens, say 300 mm, the height of the slits 10 mm and their width 0·25 mm, the distance between S_1 and S_2 100 mm, and the radiation output of the lamp 25 W, confined to a lamp aperture of cross-section of 12·5 mm × 20 mm. Using these dimensions, it is possible to estimate roughly the number of photons per second that arrive to form the interference pattern.

The fraction of the radiation from the lamp that reaches the lens and therefore S_1, is the solid angle subtended by the lens at the source, divided by 4π—that is, approximately
(area of the lens)/(area of a sphere of radius 300 mm),
which works out to

$$\frac{\pi \times 50 \times 50}{4\pi \times 300 \times 300} = \frac{25}{3600} \approx 7 \times 10^{-3}.$$

But only part of the image of the source (and so only part of this radiation) falls on the open part of the slit S_1, and therefore the reduction due to this stage is

$$\frac{\text{area of slit}}{\text{area of image}} = \frac{0\cdot25 \times 10}{20 \times 12\cdot5} \approx 10^{-2}$$

(as image is same size as object).

Assuming that the radiation through the fine slit S_1 is spread uniformly through a half-cylinder of radius S_1S_2 and height 10 cm, the reduction at the slit S_2 is

$$\frac{\text{area of slit } S_2}{\text{area of half-cylinder of radius } S_1S_2} = \frac{0\cdot25 \times 10}{\pi \times 100 \times 10} \approx 8 \times 10^{-4}.$$

Multiplying all these reductions together, and then dividing by 2 to allow for two slits, gives a total reduction of

$$\frac{7 \times 10^{-3} \times 10^{-2} \times 8 \times 10^{-4}}{2} \approx 3 \times 10^{-8}.$$

A 25 watt lamp provides 25 joule in 1 second, and so emits about $25/(3 \times 10^{-19})$ photons per second. As these are reduced by a factor of 3×10^{-8} before reaching the interference pattern, the number of photons arriving per second in the interference pattern is 25×10^{11} s^{-1}, say 2×10^{12} photons per second.

6.3. *Wave-particle duality*

If the waves coming from different parts of the light source are continuous and in phase with one another, the source is said to be coherent in space and time. This is a simple definition of coherence; perfectly coherent light is the light from such a source. In the case of incoherent light there would be a wide variety of waves leaving the source, and the waves from one point would bear no precise phase relation to those from another point, and also the waves from each point would be continuous for only a very short time. Ordinary sources are necessarily incoherent, for the time interval between photon emissions is random. It is not even the same for consecutive photons. Also the act of emission of a photon is a process that occupies a time of the order of 10^{-8} s.

The wave and particle pictures of radiation are reconciled in the idea of a *wave packet* (fig. 5.1 p. 74). The waves are there, accounting for the wave properties; and the particle properties occur because of the finite length of the packet. At first sight, this seems to emphasize the wave properties, but, a look at some typical orders of magnitude provides reassurance. The dotted line represents the envelope term defining the growth and decline of the wave packet, and for wave packets emitted by sodium lamps, the time for such a packet to pass a point in space is approximately 10^{-8} s; the length occupied by such a packet is approximately 3 m. More than a million wavelengths occur in the one packet. The average length of such a packet is known as the *coherence length* of the source, and the time taken to travel through a distance equal to this average length is the *coherence time*. These are averages, and wave packet lengths, even from the same source, vary. A perfectly coherent source would have an infinite coherence length. (See pages 96 and 97).

6.4. *Heisenberg's uncertainty principle*

The principle may be expressed in the form $\Delta p \times \Delta x \geqslant h$, where Δp is the uncertainty with which the value of the momentum of a particle may be found, and Δx the uncertainty in locating its position simultaneously.

The principle shows why it is impossible to say whether a quantum of energy is a wave or a particle, since a definite statement either way implies certainty in p or x which give a product $\Delta p \times \Delta x$ less than h.

The more photons per second that there are contributing to a pattern, the greater is the likelihood of the calculated pattern resulting. We estimated the number of photons per second in a typical Young's slits pattern as 2×10^{12}, which is so large that the conventional wave pattern and the total distribution of the individual photons are indistinguishable. This is illustrated by figs. 6.5 (*a*), (*b*) and (*c*) which are the distributions of 10, 100, and 1000 photons in part of the field of view for a Young's slits experiment. Fig. 6.5 (*a*) gives little clue to this ultimate pattern, but it is already apparent in fig. 6.5 (*c*) what form this pattern will take even though the number of photons needed to complete the pattern is 10^9 times more.

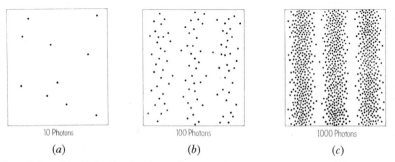

Fig. 6.5. Typical distribution of photons in two-slit interference experiment, when the number of photons is (*a*) 10, (*b*) 100, (*c*) 1000.

Similar arguments must also apply to the consideration of the paths of individual photons from the source to the screen. It would be interesting to know whether an individual photon in the situation of fig. 6.5 (a) passes through slit S_2 or S_3 or both. The uncertainty principle shows that there will never be an answer to this question. In an experiment that might enable an individual photon to be identified, the reaction of the measuring apparatus would be sufficient to destroy the wave pattern. The most that can be said about a photon is that it has travelled in some way from slit S_1 to the screen MN, and it cannot be said whether it has travelled through one slit or the other or through both of them. The uncertainty principle is *not* a confession of an ignorance which may one day be relieved. It is one of the fundamental principles of physics.

The uncertainty principle implies that the occurrence of a particular physical situation is a matter of chance. Wave mechanics is concerned with the mathematical form of this chance and how different probabilities may be calculated. In this field, the word probability has a precise meaning which lies beyond the scope of this book, but it is reasonable to talk in terms of there being a probability for each situation which can be stated mathematically. For a large number of photons there is a

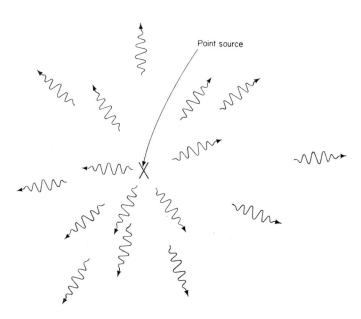

Fig. 6.6. Photon emission.

probable distribution which is more precisely calculable than the probable distribution of a small number of photons. For a sufficiently large number of photons the element of doubt is negligible, e.g. fig. 6.6 represents a small number of photons emitted by a point source. These appear to be random but nevertheless satisfy the probability distribution for a small number of photons as do the photons of fig. 6.5 (a). The random nature begins to disappear with a larger number of photons as in fig. 6.5 (c) and disappears entirely when dealing with 10^{12} photons per second, which is the number of photons in the earlier ' thought ' experiment. For a typical practical source the number of photons would be so large that space could not be found for them on a diagram of the type shown in fig. 6.6. Because of the large number of photons the random nature is lost and the probable wave pattern appears, which is the secondary wavelets pattern of Huyghens' principle, or nearly so. The remaining difference is that the interference conditions must take into account the average length of the photons (the wave packet length) emitted by the source. The wave pattern is best considered as a set of wave packets moving away from the source with time.

Application of the uncertainty principle to energy levels means that E_3 and E_2 (figs. 5.9 and 5.10) are likely to lie within a small range of values around mean values E_3 and E_2. Therefore the energy transition $E_3 - E_2 = h\nu$ is a jump from an energy band E_3 to an energy band E_2 producing a range of frequencies in the neighbourhood of ν. Similarly other transitions will provide other ranges of frequency. Figure 6.7 shows a typical distribution of displacement with frequency for one transition in a light source. All the displacements due to these frequencies may be added according to the Principle of Superposition, using the process known as Fourier synthesis. The way in which the resultant displacement varies with time is shown in fig. 6.8. This resultant turns out to be the same as a single wave train; its length is the average wave packet length.

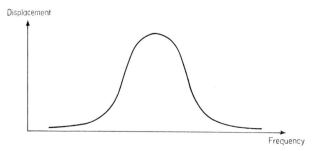

Fig. 6.7.

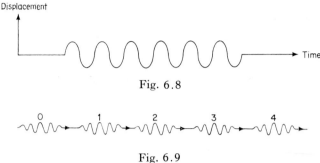

Fig. 6.8

Fig. 6.9

Figure 6.9 represents the probable positions of an individual average wave packet at times 0, 1, 2, 3, and 4 units as it moves in a straight line in one dimension. From a point source in two-dimensional space, the probable positions of the wave packets move radially outwards; for example, the probable paths to slits S_2 and S_3 are shown in fig. 6.10. After passing through S_2 and S_3, the function defining the probable wave packet position spreads in all directions, and the probable positions of the wave packets at time units 2 and 3 are represented by the arcs centred on S_2 and S_3. To find the effect at some point P on the screen

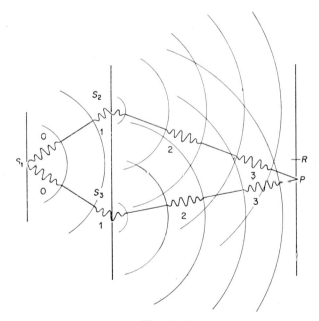

Fig. 6.10.

of many photons the total sum of the probabilities of the photons for the two possible paths is calculated. It is assumed that the wave packet probabilities leave the source in phase even if they travel in different directions. It is to be remembered that these summations are probabilities and no conclusion may be drawn about the path of one particular photon. Because the average probable position and thus the wave pattern is shown to be moving radially, it is wrong to interpret this as meaning that an individual photon itself is spread out radially. The uncertainty principle indicates that *individual* photon paths are not known and one can only draw conclusions about probabilities.

If the difference in the paths through S_2 and S_3 is very small, the summation of the average wave packets through S_2 and S_3 will be of the form shown in fig. 6.11, for brightness and darkness respectively. (There are really about a million wavelengths, represented by the four in the diagram.)

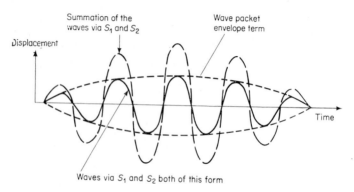

(a) Brightness

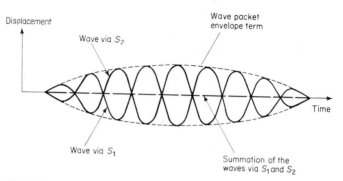

(b) Darkness

Fig. 6.11.

If the path difference to a point on the screen is greater than the length of the average wave packet, for example the point Q in fig. 6.12, then the average wave packet emitted at time t and passing through slit S_3 has reached and passed Q on the screen before the average wave packet emitted at time t and passing through slit S_2 gets to Q. Thus the summation of these two probabilities appears as in fig. 6.13. The probability due to the path through S_3 is seen to lead the probability through the slit S_2, and there is no interference between the photons arriving via each of the paths.

Why is attention confined to the average wave packet length due to emission over a small length of time? With a point source, the average

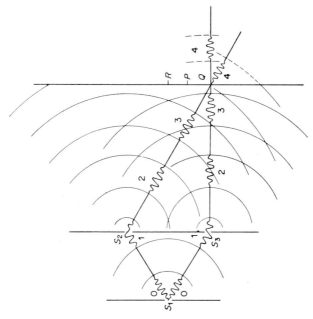

Fig. 6.12. Photon and wave front picture.

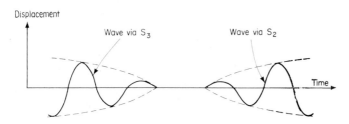

Fig. 6.13.

wave packet length at a given instant is the same in all directions; but it will vary with time since the physical conditions of the source will vary, though not by much. However, this variation is sufficient to alter the interference conditions. For example, a fluctuation of one part in a million in the energy could mean a change of one part in a million in the average wavelength. The summation of the photons may be considered equivalent to the summation of the two *average wave packets.* Consider a point Z such that an average wave packet of one million wavelengths arrives at the same time as an average wave packet produced after an energy change of one part in a million and coming by a different path. At Z the following sequence of events would occur. Initially the two average wave packets would arrive in phase. They would gradually get out of step until they are completely out of phase as the mid-points of the average wave packets pass Z. Then they gradually get in step again until they are in phase as the average wave packets leave Z. This all takes place in about 10^{-8} s, and the interference pattern changes far too rapidly to be observable. Interference only takes place between a large number of photons, all of which are emitted within a short time interval, the time taken for the average wave packet due to these photons to pass a point in space. If the optical path difference is greater than the average wave packet length (which is called the coherence length), then there is no interference.

Now that the importance of coherence has been established, the factors that determine its length are considered. The graph of displacement against time of fig. 6.8 is equivalent to the graph of displacement against frequency of fig. 6.7. The longer the wave packet the narrower the frequency range. An infinitely long wave packet would give an infinitely narrow frequency range, which would mean a perfectly pure spectral line. The converse Fourier synthesis is also true, and a range of frequencies must be equivalent to a single wave packet of finite length. The wider the range of frequencies (usually called the line width) the shorter is the length of the wave packet. The line width is inversely proportional to the coherence length. The energy of a given excited state can vary slightly about a mean. The range of possible energies has an associated range of frequencies, which (translated into wavelengths) is known as the natural line width for the transition concerned. This line width has a corresponding coherence length whose value for the 589 nm sodium line is calculated to be about 3 m.

The range of frequencies is increased by the Doppler effect due to the speed of thermal motion of the atoms (the main effect) and also by collision with other atoms (pressure broadening). As a consequence of this there is a decrease in coherence length to about 10 cm; this value depends on the number of atoms present per unit volume (the pressure), and the temperature of the emitting gas. There is a further reduction in coherence length in a sodium lamp, which is due to the variation in conditions across the width of the emitting parts of the lamp. The effect of this reduction may be lessened by focusing the light from the source on to a pin-hole which may be considered to act as a Huyghens' secondary source emitting light that is dependent on the sum of all the light focused by the lens at the pin-hole immediately before re-radiation. This sum may vary with time but the light emitted from a very fine pin-hole will remain constant for the time taken to emit one average wave packet. If the pin-hole is fine enough to produce this improvement, then there is a considerable reduction in the light intensity. The coherence length for the light from an arrangement which starts with a sodium lamp will then remain at about 10 cm. Without the fine pin-hole the coherence length for a sodium lamp is approximately 1 cm.

The line width, and hence the coherence length of a laser source, depends in a fairly complex way on several experimental situations. The first to be considered is the effect of the amplifying stage of laser action on the natural line width. Because stimulated emission is the cause of this amplification, and this stimulation is a resonance effect between a stimulating photon and an excited atom, the stimulated photon is in phase with the incident photon. The effect is to increase the length of the wave packet associated with the incident photon. This increase has in certain (rather special) circumstances been as high as hundreds of kilometres. Most practical lasers have coherence lengths of the order of 40 cm. This by itself does not appear to be much of an advantage over the conventional sodium lamp. However, there are two other points to be considered. First, the intensity of the sodium lamp has to be reduced considerably to achieve this level of coherence, and in these circumstances should really be compared with a laser operating under very special conditions, which would be found to have a much higher coherence length. Secondly, there is a much better average wave packet correlation across a laser beam than there is across a sodium discharge lamp. The laser needs no pin-holes or apertures to restrict the beam to a coherent part of itself, so that the coherent output is more intense.

CHAPTER 7
resonance and line width

7.1. *Mechanical demonstration of line width*

THE principles of resonance and line width have a definite role to play in laser design and are therefore considered next. Figure 7.1 illustrates a well-known mechanical demonstration of resonance. One end of a wire is attached to a vibrator and the other end to a mass which hangs over a pulley at the end of the bench. The vibrator is excited by an alternating current from a signal generator. Suitable values might be a tension of about 0·7 N produced by a mass of 0·07 kg, and a 2 m length of 38 s.w.g. copper wire. If the signal generator is switched on at a low frequency, say, 15 Hz, the wire is unaffected, apart from small forced vibrations close to the vibrator. As the frequency is steadily increased a situation arises when the wire is seen to vibrate strongly at right angles to its length, according to the equation, $f = 1/2l \sqrt{(T/m)}$, where f is the frequency of vibration, l is the length of wire, T the tension in the wire and m the mass per unit length. If the frequency is further increased slowly, the amplitude of the vibration will increase to a maximum and then fall off as the frequency is further increased. The frequency which gives the maximum amplitude is called the resonant frequency for the fundamental, or first harmonic. Note that the vibrator will excite the wire quite strongly for a range of frequencies, of the order of 4 Hz, on either side of the actual resonant frequency. This range may be examined in detail using a travelling microscope to obtain

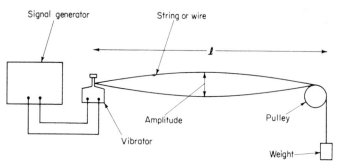

Fig. 7.1. Stationary waves on a stretched string.

the mid-point amplitude at different exciting frequencies. All the observations should be taken fairly quickly, since the signal generator must be stable to about 0·05 Hz. An eyepiece using a graticule rather than cross-wires helps, and so does good illumination of the wire.

The observations are shown as a graph of frequency against (amplitude)^2 (fig. 7.2). Different samples of wire give different graphs (fig. 7.3), but these graphs are all of the same form, which is, at first, each falling fairly steeply on either side of the maximum, and then after points B and E (fig. 7.2) fading more slowly to zero. It is impossible to specify exactly the range of frequencies which can keep the wire vibrating at its fundamental frequency, so a new idea, that of line width is introduced. If the frequency scales of fig. 7.3 (*a*) and (*c*) are altered they may be made to look like fig. 7.3 (*b*), as shown in fig. 7.4 (*a*), (*b*) and (*c*). Therefore *all these curves are a kind of ' line '*. This ' line ' is not the perfect line of geometry books, which has position with no thickness. The lines of fig. 7.4 (*a*), (*b*) and (*c*) have a measurable thickness and a very definite profile or shape. The thickness is different for different amplitudes, so the convention has been adopted that the *width* of the line is measured where the (amplitude)^2 is half the maximum (amplitude)^2, XY in figs. 7.3 (*a*), (*b*) and (*c*). This is the line width, and it has the values of 2·5 Hz, 1 Hz and 6 Hz, for the cases (*a*), (*b*) and (*c*) respectively.

With regard to terminology, in optics, the word 'line' means an optical spectrum line of definite frequency. Here the idea of ' line ' has been reached via a mechanical counterpart which reproduces the optical effect to scale.

These differently shaped curves correspond to different degrees of damping or dissipation of energy. In situation (*b*) less energy

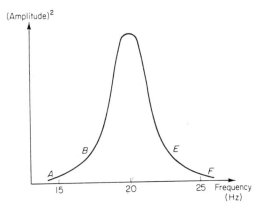

Fig. 7.2.

is lost per cycle than in (a), which in turn has less energy lost per cycle than in (c).

Now, the narrower a line is for its height, the more it resembles a line with no thickness—and so in a sense the higher is its 'quality'. Quantitatively, the quality factor Q for an oscillating system can be defined as

$$Q = 2\pi \times \sqrt{\left(\frac{\text{energy stored per cycle}}{\text{energy dissipated per cycle}}\right)}$$

It can be proved that $\nu_0 = Q \times \Delta_0$, where ν_0 is the resonant frequency and Δ_0 is the line width. So for a given frequency, narrow line width corresponds to a large Q factor. Laser resonators belong to this category because the output from a laser is spread over a very narrow frequency range.

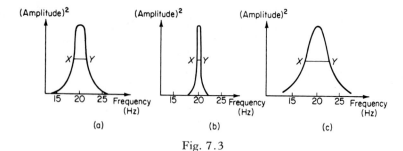

Fig. 7.3

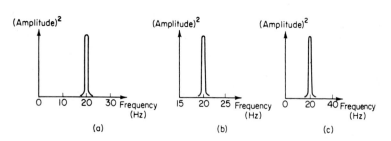

Fig. 7.4.

Returning to the mechanical experiment, if the frequency is further increased steadily, additional resonances corresponding to new standing wave patterns will be obtained as fig. 7.5 the frequencies $2\nu_0, 3\nu_0, 4\nu_0, \ldots$ of the second, third, fourth, and higher harmonics are reached, perhaps up to the 20th harmonic (for which the wire is 10 wavelengths long).

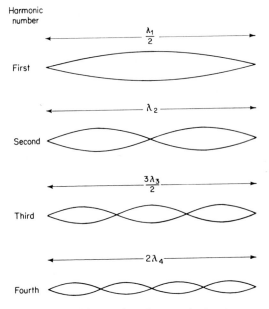

Fig. 7.5. Harmonics of a stretched string.

7.2. Resonant cavities

In Chapter 5 (pp. 84, 85) reference was made to the need to place mirrors at either end of lasers to reflect the laser light back and forth to produce sufficient stimulated emission. The space between the mirrors is known as the laser cavity and it is normally long and narrow. For a helium–neon (He–Ne) gas laser typical dimensions are 300 mm length and 10 mm width (fig. 7.6). The number of complete wavelengths between the mirrors for light of wavelength 633 nm is equal to $(0\cdot3/633) \times 10^9 \approx 10^6$. The cavity for the first He–Ne laser had parallel polished plane mirrors and was like a Fabry–Pérot interferometer. This sort of cavity is called a Fabry–Pérot resonant cavity (fig. 7.7 (a)).

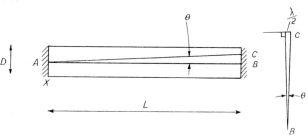

Fig. 7.6. Resonant cavities.

Suppose that the distance between the plane mirrors for fig. 7.6 is an exact number (n) of wavelengths. Consider what happens to light starting along a path AC, at an angle θ to the axis AB where θ is chosen so that $AC - AB = \lambda/2$. This means that back and forth once in the general direction AC produces a path of length $n+1$ wavelengths. If the diameter of the mirror is D and the length of the cavity is L, then since θ is a small angle, BC may be taken to be equal to $L\theta$, and BC. θ equal to $\lambda/2$. Therefore the maximum number of reflections at angle $\theta = D/2L\theta$, and $L\theta^2 = \lambda/2$ and $\theta = \sqrt{(\lambda/2L)}$. Therefore the maximum number of reflections at angle $\theta = D/\sqrt{(2L\lambda)}$. For the typical dimensions of a cavity referred to earlier the number of reflections $= 10 \times 10^{-3}/\sqrt{\{2 \times 0 \cdot 3 \times 633 \times 10^{-9}\}}$, which is approximately 17 reflections and is sufficient to support laser action provided other operating conditions are satisfied.

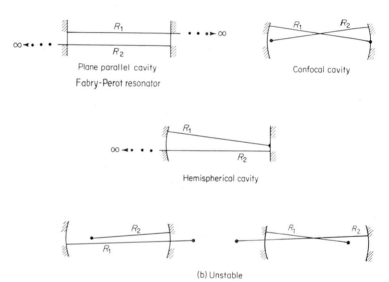

Fig. 7.7. Fabry-Pérot Resonator.

7.3. Multimode laser beams

For greater lengths this number reduces, hence the shorter the gas laser the greater the likelihood of multiple resonant positions being observed. If the beam of such a laser is allowed to fall onto a screen about 10 m away, then it will appear to have a patterned structure (fig. 7.8). Case (a) is called single mode or uni-phase operation; all points in the field marked with the arrow are in the same phase. In case (b) the phase in each section is π out of phase with its neighbour, as shown

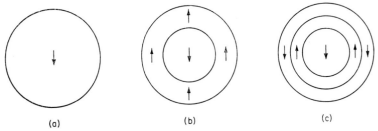

Fig. 7.8. Modes of resonant cavities.

by opposite arrows. Where the modes overlap they interfere destructively, giving a dark ring pattern. The centre of the whole pattern will be more intense, since obliquely travelling light loses energy through the sides of the tube. If the side losses for a section or mode are greater than the average gain due to stimulated emission, then this outer section or mode will not appear. For a given tube, the lower the input the smaller is the number of modes that can be supported, and geometry shows that the narrower the tube the smaller the number of modes present. These are the two simplest recipes for removing the additional modes; the first may be partially in the user's control but the second is decided by the manufacturer.

If the mirrors were at an angle to each other, further calculations would show that other mode patterns, (fig. 7.9) appear. Thus any slight tilt in the mirrors changes the mode pattern of the beam, and these changes occur very easily for the arrangement of parallel mirrors, so much so that lasers using this design have to be adjusted frequently during operation. This situation may be considerably improved, but not eliminated, by replacing the plane mirrors with concave mirrors.

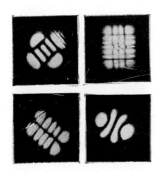

Fig. 7.9. Laser mode patterns.

Many arrangements of mirrors have been used; a selection is shown in fig. 7.7. One of these, the hemispherical cavity, produces maximum output in a single fundamental mode and is relatively insensitive to mechanical changes. The difference between the stable and unstable systems is that the very slight misalignment with unstable systems is sufficient to stop laser action, but for stable systems equivalent misalignment only produces a reduction in the intensity of the output. If a plane mirror is used, it must lie at or inside the centre of curvature of the concave mirror for the system to be stable, otherwise the plane mirror will reflect light along a path that lies outside the perimeter of the concave surface.

7.4. Interference mirrors

Whilst discussing mirrors it is appropriate to mention their coatings which are especially made to provide the high reflection coefficients necessary for laser action. The mirrors are made from an odd number of layers of dielectric media. Each layer is one-quarter of a wavelength thick and each alternate layer has an electrical permittivity (and hence optical refractive index) whose value lies between those of air and glass; the other alternative layers have an electrical permittivity that is greater than that for glass. Such a system provides destructive interference for waves following a transmission path, and constructive interference for waves following a reflection path, thus producing a coefficient of reflection which is close to 100 per cent. The more layers present, the higher the coefficient of reflection. However, since this phenomenon is based on thicknesses equal to an integral number of quarter wavelengths, each mirror is made specifically for the predominant wavelength required from the laser, and such mirrors are not capable of providing a high coefficient of reflection for a wide range of wavelength.

For example, for the He–Ne laser two wavelengths of particular interest are 632·8 and 1152·3 nm. For a mirror chosen to suit *one* of these wavelengths, the reflection of the *other* wavelength is so small that laser action cannot be supported, or if it is supported, it will be very weak indeed.

The mirrors may be aligned by adjusting the spring mountings, and it will be noticed that if one of the mirrors is tilted while the laser is working, the mode pattern changes. If the beam is projected onto a plane mirror some distance away and reflected back on to a screen alongside the laser, the patterns can be seen clearly while the screens are being adjusted, and one should be able to concentrate the energy into a single mode, or at least into a small number of modes. Physical optics experiments in general, and of course holography, demand single mode (uniphase) operation, otherwise each mode will act as an independent source and there will be several patterns instead of just the one required.

7.5. *Laser line width*

In a laser there are several resonance actions and the resultant beam is the summation of these actions.

The energy transition from state E_3 to E_2 has a spread of frequencies as mentioned on page 79 and this spread is shown in fig. 7.10. This is the natural line width for spontaneous emission at this frequency range. Stimulated emission for the transition E_3 to E_2 produces a much narrower line width and is shown as the laser line width in fig. 7.11, which is drawn on the same scale as fig. 7.10.

The laser line width range is increased by Doppler effect and pressure broadening as mentioned on page 96. This new range of frequencies

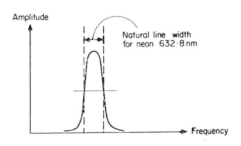

Fig. 7.10. Frequency spread of laser beam.

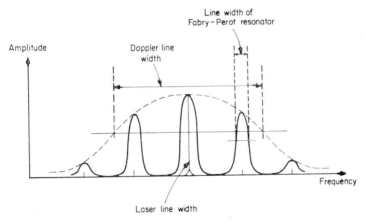

Fig. 7.11. Resonator line widths.

is shown as a broken-line envelope in fig. 7.11, and it is much broader than the laser line width.

The laser's resonant cavity has a line width of its own which is less than that for the Doppler effect, etc., and means that the cavity is selective in supporting only some of the available frequencies around a mean frequency that corresponds to the whole number wavelength condition evaluated on page 101. In that discussion it was shown that there might be more than one whole number wavelength and corresponding mean frequency capable of supporting laser action which is why several resonator line widths appear in fig. 7.11. The possible laser output is the sum of these two effects and is shown by the heavier continuous line in fig. 7.11. Whether all these frequencies appear in the final output depends mainly on the amount of energy put into the excitation process. The gain due to stimulated emission depends on the number of excited atoms present and could be indicated by a horizontal line in fig. 7.11. All parts of the frequency range above this horizontal line would appear in the final laser output. If there is more than one peak above the line then the laser would be operating in a multimode fashion.

CHAPTER 8
laser construction

8.1. *The ammonia maser*

THE first successful device using stimulated emission was the ammonia maser designed by C. H. Townes and his co-workers at the Columbia University, U.S.A. This device operated in the microwave region of the electromagnetic spectrum. The word maser stands for Microwave Amplification by the Stimulated Emission of Radiation.

In fig. 8.1 the black dots represent the positions of the three hydrogen atoms, and the circles 1 and 2 are the two possible positions for the one nitrogen atom. These two positions of the nitrogen atoms give two different energy states. If the nitrogen atom is at 1, this is the slightly higher energy state, which is called the excited state; if it is at position 2; it is at the lower, unexcited, state. The difference in energies of these two states corresponds to a frequency of 23·9 GHz.

Molecules in both states occur naturally and some ammonia gas is kept in an 'oven' (fig. 8.2), whose temperature and pressure are

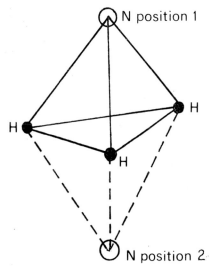

Fig. 8.1. Structure of ammonia molecule.

controlled. Both types of molecules are allowed to escape through an aperture A. The difference between these two energy states is increased in high electric fields, of the order of 10^6 V m^{-1}. This property is used to separate the two types of molecule by passing them through an inhomogeneous electric field. The excited molecules are focused into a beam by the field, and the unexcited molecules are scattered off-axis. The beam of excited molecules is then passed through an aperture into a hollow chamber (resonant cavity) which is tuned to resonate at the microwave frequency 23·9 GHz.

Early masers were used as amplifiers of external signals and the cavity was supplied with both input and output waveguides. The population of excited molecules was adjusted so that no stimulated emission was observed until a small input signal was fed into the cavity.

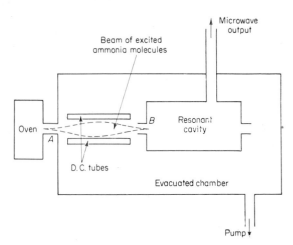

Fig. 8.2. Schematic diagram of ammonia maser.

8.2. *The ruby laser*

After the invention of the ammonia maser the race to produce the first optical maser was on. It was won by T. H. Maiman in 1960 at the Research Laboratories of the Hughes Aircraft Co. in California.

He used a specially grown crystal of ruby in the shape of a cylinder with highly polished ends. Ruby is a crystalline form of aluminium oxide, and its characteristic pink colour is due to a 0·05 per cent 'impurity' of chromium ions regularly spaced in the aluminium oxide crystal lattice. Suitable ruby crystals have a diameter in the range 1 mm to 25 mm, and length in the range 20 mm to 400 mm. The crystal is mounted co-axially within a helical xenon flash tube (fig. 8.3) which is operated at about 5000 V. The flash tube and ruby crystal are themselves mounted co-axially inside a cylindrical container with highly polished walls. The tube is flashed at intervals, by discharging a bank of large capacitors. One of the polished ends of the ruby crystal is coated with a 'semi-reflecting mirror', which in fact reflects about 85 per cent of the light. The other polished end is coated with a mirror which reflects as perfectly as possible. The ends of the crystal are parallel (within something closer than 10 seconds of arc) and this property is a most critical one for good performance.

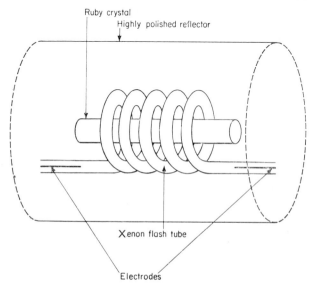

Fig. 8.3. Scheme of ruby laser.

When the xenon tube flashes, the crystal is exposed to a total radiant energy of about 1000 joule over a period of a few milliseconds. Much of this energy is in the blue and green part of the spectrum. Some of the energy is absorbed by the chromium ions of the ruby crystal, in strong absorption bands which are known as $^4F_1$ and $^4F_2$ in spectral notation (fig. 8.4). This absorption, and the breadth of the absorption band, leads to a considerable population surplus in the $^4F_2$ band. These excited ions become less excited by falling to a lower energy state, the ^2E state; the small amounts of energy released by this transition are stored as lattice vibrations. The state ^2E is a *metastable state*, which means that it has a much longer average lifetime than a normal excited state. The possible changes of energy state are summarized by certain 'selection rules'. These rules show that, since an electron can only go to a state where there is a place available for it, certain transitions are considered to be impossible. These are called 'forbidden' transitions. So if an atom somehow reaches an excited state in which direct return to the ground state is forbidden, with no intermediate energy state in between, it ought to have to stay in this excited state. Quantum mechanics produces the same idea, but shows that the 'forbidden' transitions can happen, but are just not very probable. Such an excited state is therefore not permanent but is *metastable*. As the probability of transition is low, excited ions accumulate in the metastable state, producing a *population inversion*. When one of these low-probability transitions does take place, a photon corresponding to the jump from state ^2E to $^4A_2$ (the ground state) is emitted; the wavelength is 694·3 nm ('ruby red'). This photon travels along the tube colliding with other metastable excited atoms, and stimulates these in turn to produce photons of the same energy. One such photon travelling along the axis of the crystal can be reflected back along its own path by the mirrors at the end faces. It is continually being reinforced coherently by the photons whose emission it stimulates. A photon which does not travel axially escapes

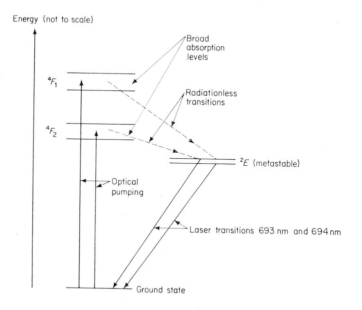

Fig. 8.4. Ruby laser action; the levels in the chromium atom that take part.

through the walls of the crystal before it is much reinforced. So most of the stimulated emission takes place axially, giving the intense coherent beam characteristic of the laser.

Other possible laser transitions in ruby may be produced by altering the concentrations of the chromium ions or the temperature of the crystal, but the transition just described is the easiest one to stimulate. The whole action takes place rapidly in a matter of milliseconds, and all the time-references in the above section are only relative to one another. Maiman's flash took approximately 0·5 ms to build up a sufficient population inversion before coherent radiation was observed. In something less than a further millisecond, the laser action ceased. The tube was then flashed again and this cycle repeated. The frequency of repetition of the flashes is controlled by the auxiliary circuit discharging the capacitors. Two factors have to be borne in mind. Firstly, the tremendous heat produced by such a flash; in Maiman's tube the input to the flash was 1000 J, and the laser output was about 1 J; most of the difference appeared as heat. If the flashes are too frequent, the system becomes overheated and the crystal may crack. Secondly, it takes a definite time to build up a population inversion sufficient to reach the 'laser threshold', which is the stage at which the gain of photons by stimulated emission just balances the losses at the mirrors, walls and elsewhere. The time needed to build up to the threshold is the reason why the output from a ruby laser appears in the form of pulses. Many of the lasers that employ solids are *pulsed lasers* of this type.

There is no coherence between consecutive pulses; coherence is confined to the very large number of photons within each pulse.

Considering the output from Maiman's ruby laser to be 1 J, and taking the energy of one photon to be approximately 3×10^{-19} J, then the laser pulse consists of around 10^{19} photons, so the coherence of the pulse within itself is significant.

At first sight, pulsing appears to be a tremendous disadvantage. This is not necessarily so, since for many uses coherence of the whole beam is not essential.

Why bother to pulse the flash tube, and why not use an ordinary discharge lamp rather than a xenon flash lamp? The answer lies in the order of energy required. To provide the population inversion that laser action demands, a certain minimum energy per unit volume must be supplied; and if it is supplied *in this way* the energy must all be of wavelengths that corresponds to the transition from the ground state to the levels $^4F_1$ and $^4F_2$. Such large quantities are only obtained easily from flash tubes. Since the early lasers of the sixties, there has been much development. There are now continuous-wave ruby lasers which are not pulsed. However, their outputs, whilst being coherent for long periods of operation, are of much lower power; a single 1 ms pulse from a pulsed laser carries more than a thousand times the energy of a continuous-wave laser that is operating for 1 ms. The repetition rate of the pulses for a pulsed laser can now be quite high as well, of the order of 10 000 per second. But the operating conditions for continuous-wave crystal lasers are considerably more exacting than those for pulsed lasers. For example, the crystal has to be cooled by immersion in liquid nitrogen (77 K).

8.3. *Q-switching*

The output of a laser pulse may be increased by a controlled delay in the build-up process. This is known as ' Q-switching '. In this situation the ends of the ruby crystal are not coated. Mirrors placed axially some distance beyond the ends of the crystal provide the necessary repeated reflection of the light. Between one of the mirrors and the crystal a fast light-switch, such as a Kerr cell, is placed. An alternative arrangement is to have one of the end mirrors rotating rapidly about an axis at right angles to the crystal axis. In both arrangements the effect is to build up a higher population inversion in the laser cavity, which is suddenly released as stimulated emission when the light-switch is opened. With the Kerr cell, this opening is done by switching off the cell's ' shutter ' electric field so that it transmits light. The rotating mirror ' switches ' the light through the laser cavity when it is in a position parallel to the other mirror.

The operational sequence of events is as follows. The electric field is applied across the Kerr cell; the tube flashes and after at least 0·5 ms the electric field is switched off as suddenly as possible. Within 0·2 ms the laser flashes with considerably increased power. The sequence for the rotating mirror is similar but the switching takes longer. Mechanical switches are slower than electronic switches. The original experiments in this direction produced an increase in peak intensity from 6 kW to 600 kW peak intensity. This is not an increase in total energy of the pulse but an increase in the peak intensity; in fact the overall energy turns out to be less. The switch lowers the amplification rate so that in the crystal there is a greater build-up of energy which is suddenly released when the switch is opened. All the switch does is to raise the threshold level. This additional peak intensity is extremely useful in itself, but other properties are also interesting. Firstly, the opening of the switch may be timed more precisely than the beginning of a conventional pulse, for which the flash tube may be fired accurately but the delay before pulsing varies somewhat by fractions of a millisecond; a Kerr cell may be operated to within fractions of a *microsecond*. Secondly, the quality of the shape of the pulse is improved, as will be seen by comparing figs. 8.5 (*a*) and (*b*).

The flash tube does not have to be helical. Another popular design in use has a reflecting tube of elliptical cross-section surrounding the crystal and flash tube. The crystal is placed axially along a line determined by one focus of the ellipse, and the flash tube, in this case now straight, is placed axially along a line determined by the other focus. The geometry of the ellipse, fig. 8.6 is such that angles such as ABC and CBD are equal, where A and D are foci, B is any point on the ellipse, and BC is the normal to the ellipse at B. This means that if the flash tube is

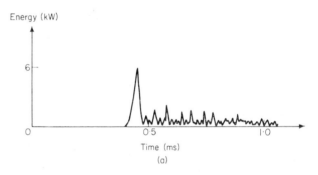

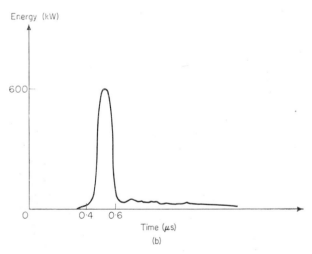

Fig. 8.5. Pulsed power output.

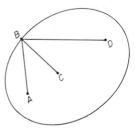

Fig. 8.6.

at A and the crystal is at D, then light reflected at any part of the wall must pass through the crystal. There are other designs of reflectors and tubes, but the guiding principle is always the same, to obtain in the crystal as large a quantity of light of the appropriate frequency in as short a time as possible.

8.4. *Other solid state lasers*

Any crystal type laser is known as a solid state laser. The range of materials so used has expanded greatly since Maiman's original discovery. The list of suitable materials grows larger every year. Laser action has been observed with nickel ions in magnesium fluoride, with many of the trivalent lanthanide ions in various host materials, and with trivalent uranium ion in calcium fluoride.

Neodymium in glass, which strictly is not a crystalline material, is also used quite extensively. Glass has the advantage that it is more easily prepared than crystalline substances whose crystals have to be specially grown with the correct degree of active material in the host material. However, although the list of possible solid state laser materials is now quite extensive, the number commercially available is still rather limited, with ruby being the most competitive in terms of cost.

8.5. *The helium–neon laser.*

Gas lasers give a continuous-wave output, whatever the gas used. The first successful gas laser employed a mixture of helium and neon. It was made in 1961 by three workers at the Bell Telephone Laboratories in the U.S.A., A. Javan, W. R. Bennett, Jr., and D. R. Herriott. Their apparatus is represented in fig. 8.7. The central tube is about 100 cm long and 1·5 cm internal diameter The. plane mirrors on the ends are mounted so that they may be adjusted parallel to each other, to better than 10 seconds of arc. The bellows enable this to be done while the tube remains uncontaminated by air. The tube was first evacuated and then filled with a mixture of helium and neon, the partial pressures being 8 mmHg helium and 1 mmHg neon. The proportion 8 : 1 is not critical, and anything between 10 : 1 and 4 : 1 is suitable. Excitation is done electrically instead of the optical method of the ruby.

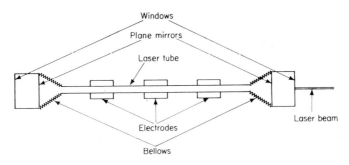

Fig. 8.7. Scheme of helium–neon laser.

In the original experiment a 50 W radio-frequency signal, in the range 20–30 MHz, was applied to the electrodes surrounding the tube. This high frequency field excites the helium atoms to energy states above the ground state. The excited atom emits a photon and in so doing reverts to the ground state. This is a continuous process as long as the field is applied.

Many sets of transitions can and do occur in this situation. Each transition produces its corresponding energy release in the form of radiation of a precise wavelength, which is observed as a spectrum line. In terms of laser action, one particular sequence of steps in the loss of energy from the excited atoms is important. Spectroscopists' notation for labelling different energy states is used in fig. 8.8. It is not proposed to explain this notation only to use it as a means of identification. The energy states concerned are marked on fig. 8.8. Those for helium are shown on the left of the figure, those for neon on the right, and they are both on the same energy scale. There are other states for helium and neon but these are the principal ones concerned with this laser action.

The 3s and 2s states for neon are very close in energy to the 2^1s and 2^3s metastable states for excited helium atoms. Being metastable these states have a longer lifetime than the other excited states which means that they become relatively well-populated. For these metastable helium atoms, the most probable way of losing energy is transferring it

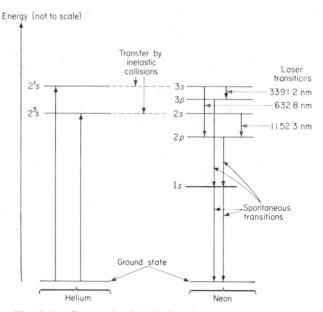

Fig. 8.8. Energy levels in helium atom and neon atom.

by ' collision ' to neon atoms. Because these excited helium states have a longer lifetime, and also because helium atoms have a preponderance of 8 to 1, the net effect on the neon atoms is population inversion between the 3s and 2p states; the 2s and 2p states; and the 3s and 3p states.

A feature of these transitions is that they end not at the ground state, but at an excited state (2p or 3p). These less excited (2p or 3p) atoms after the stimulated emission do not persist and hence oppose further laser action, but steadily and spontaneously return to lower states (1s and ground). After they have played their part they leave the stage. Population inversion is only required between the states on either side of the stimulated transition. Thus for the 632·8 nm line the required inversion is between the populations of the 3s and 2p states, and it is independent of the populations of the lower states. These population inversions are maintained continuously by virtue of the continuous helium discharge. In addition to these population inversions many other inversions may be produced with neon.

Population inversion is the first stage in laser action; the next (as with the ruby laser) is to make photons pass back and forth along the tube, accumulating by stimulated emission. The first He–Ne laser used plane mirrors, which reflected one or more chosen wavelengths much more efficiently than others. The wavelength, 632·8 nm, is a red line, and the wavelengths 1152·3 nm and 3391·2 nm are in the near infrared. The same apparatus, with the necessary interchange of mirrors, will produce all these wavelengths; the output for either of the the infrared lines will normally be greater than that of the visible line, this is especially true for the 3391·2 nm line.

8.6. Brewster windows

Windows at the end of the laser tube allow the necessary mirrors to be set outside it, an arrangement which has several advantages; the windows make for better joints than was possible using bellows. In addition, they stop gas interaction with the coatings of the mirrors during the various pumping and filling processes needed to fill the tube with the correct helium–neon mixtures. Also there is more flexibility in terms of interchange of mirrors, replacement of tubes, etc. One difficulty is that a normal glass window reflects about 8 per cent of incident light, and this reflected light would be out of phase with the light that has been transmitted through the window and been reflected by the mirror. The gain in many laser tubes is only about 4 per cent per metre length and this means that a laser with normal glass windows would produce out-of-phase photons faster than more photons would be produced by stimulated emission.

This difficulty was overcome by using Brewster windows (pp. 66–68) with their high coefficient of transmission 99·9 per cent. This high coefficient applies only to light that is plane polarized in the plane of incidence. Use of Brewster windows necessarily results in the laser light being plane polarized, for the following reasons. The first few photons of the laser amplification process are distributed randomly through various planes of polarization. Those that are polarized in the appropriate plane will pass straight through the Brewster window to the mirrors beyond, and then be reflected to and fro across the cavity, losing very little energy. A proportion of the photons that are not polarized in the appropriate plane are reflected out through the side of the laser, and the reduction by this process exceeds their multiplication by stimulated emission. Therefore, only light with a plane of polarization that is in, or very close to, the polarization plane appropriate to the Brewster angle takes part in the amplification process. The resulting beam is plane polarized to closer than 999 parts in 1000.

Figure 8.9 shows the tube of a helium–neon laser, with the Brewster windows at its ends. Not all He–Ne gas lasers use Brewster windows now.

8.7. D.C. excitation

Early gas lasers required radio-frequency excitation. At these high frequencies, the vacuum seals at the Brewster windows tended to fail. Later models used hot cathodes (fig. 8.9), the helium atoms being excited by direct current, approximately 15 mA, 2000 V. Most recent versions have reverted to cold cathode. Both anode and cathode are in side tubes so as to be out of the path of the laser beam. The population of excited atoms is controlled by the applied voltage, which in turn controls the current through the discharge. These are not linear effects, since there is a saturation current which depends on the pressure. With meters in the power supply it is observed that as the discharge current rises above the threshold needed for the laser to work at all, the intensity of the radiation rises, rapidly at first and then more slowly until an upper limit is reached. A laser has a longer useful life if it is under-run. One of the main advantages of d.c. supply is that the laser output is steadier; with care, ripple can be reduced to less than 0·1 per cent.

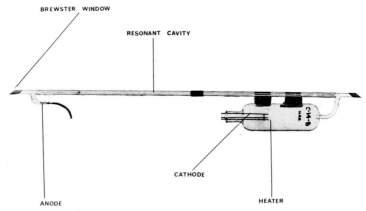

Fig. 8.9. Helium-neon laser.

8.8. Other types of gas laser

The argon-ion and krypton-ion lasers, have two advantages over the helium–neon laser. Firstly, there is a considerable increase in output. The best available helium–neon output is 75 mW uniphase, or 150 mW multimode, whereas 2 W total or 900 mW uniphase is available with argon-ion lasers. Secondly, there are many more wavelengths in the visible spectrum that can support laser action. The argon-ion laser gives eight different wavelengths all in the blue-green part of the spectrum, and the krypton-ion laser gives nine lines spread through the visible part of the spectrum.

Ion lasers are much more intricate in design. The discharge tube forms a closed rectangle, fig. 8.10, the laser capillary tube being one of the long sides. Electromagnetically, the discharge tube is equivalent to an ordinary wire loop carrying a current. This loop is coupled to a radio-frequency supply circuit and the coupled circuits may be considered as a radio-frequency transformer, a system which at the energies required saves a number of technical problems since the electrodes are external to the tube. An ionized gas is a plasma, so that we are using a plasma discharge. Direct-current discharge tubes have also been made for use with argon ions. The current densities are of the order of a few hundred amperes per square centimetre, so there is considerable heating, which means that a water-cooling jacket is needed. The intense discharge drives gas ions into the walls of the tube, so that the gas pressure drops. This was dealt with in the early tubes by pumping fresh gas at the correct pressure through the tube, a continuous process which also provided additional cooling. At some of the lower outputs now available, around 1 W, it is possible to manage with sealed tubes which have a gas reservoir in a side arm for topping up the gas pressure (reminiscent of the early X-ray tubes which met exactly the same problem and solved it in the same way).

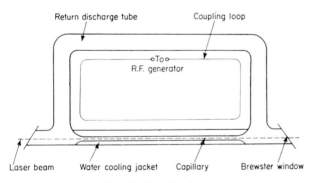

Fig. 8.10. Ion laser.

' Getters ' are mounted in the side arms for cleaning up any unwanted gases that may emerge. A solenoid surrounding the tube enables a high magnetic field to be applied axially to the plasma which increases the output in single mode operation; the flux density is of the order 0·1 tesla. Brewster windows are used as before, but one of the end mirrors is replaced by a Littrow prism, which has one face coated as a plane mirror. The light paths for blue and red light in such a prism are shown in fig. 8.11 (a) and (b). In (a) the red light returns along its own path and the blue light is refracted out of the beam. In (*b*) the incident light is in the same direction as (*a*), but the prism has been rotated anti-clockwise, and now it is the blue light that returns along its own path, while the red light is refracted out of the beam. (In these figures blue and red have been separated for clarity, but in practice they overlap each other as far as their incident paths are concerned.) Thus, by rotation of the Littrow prism, the laser may be operated at any of the wavelengths corresponding to a standard argon transition. This type of laser requires mirrors that reflect over a wide spectral range, not the multiple coated selective mirrors of the helium–neon laser with their limited frequency response. The ancillary equipment for ion lasers is more complicated than that for helium–neon lasers, but the ion lasers give much more power over a wider wavelength range in return for this increased complexity.

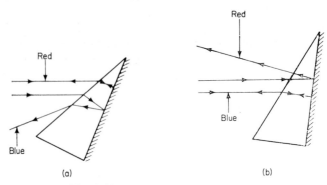

Fig. 8.11. Littrow prism end window.

Among the most powerful gas lasers nowadays are the carbon dioxide molecular gas lasers, whose main operating wavelength is 10·6 μm, (or 10 600 nm) which is in the infrared region of the spectrum. They are called molecular gas lasers, because the energy transitions involved are molecular energy changes (like that of the ammonia maser), and not electron energy changes. The principles of operation are similar to those already outlined, except that the energy changes are of the same order of magnitude as thermal energy changes, which themselves correspond to wavelengths in the middle and far infrared. So the tube has to be maintained at a steady and preferably low temperature so that ordinary heat exchange with the surroundings does not alter the molecular population inversion, and water cooling is used. Usually the mirrors are mounted directly on the ends of the laser tube; sodium chloride is, so far, the only successful Brewster window material for the infrared and sodium chloride crystals of optical quality are difficult to handle. The power output is directly proportional to the length of the laser tube. Outputs of the order of 100 W and more have been achieved. By 'folding' techniques, lengths equivalent to 200 m have been attained and the corresponding power output is of the order of kilowatts. Because of these high powers, lasers from this category are the ones, normally used for the spectacular hole burning experiments through bricks, etc. A gold coating on a stainless steel heat sink has been used for the high reflection mirror. The 'semi-reflecting' mirror is made of germanium at the time of writing.

8.9. *Semiconductor lasers*

Semiconductors are solids having an electrical conductivity which, whilst not being high, is considerably greater than that of a good insulator such as rubber. The conductivity may be altered by introducing impurities in a controlled fashion (doping). The amounts of impurity are small, up to 0·1 per cent; and in most cases much less than this. If the impurity atoms have one more electron in their outer shell than the atoms which they replace in the host lattice, then the impurities provide a surplus of electrons in the host material. Such material is known as n-type. The opposite case is when the host material is doped with atoms that have one less electron in the outer shell than the atom which is replaced. This material is known as p-type. The first semiconductor lasers employed gallium arsenide.

In a semiconductor laser, p-type material is produced alongside n-type material in a crystal, forming a p–n junction at the boundary of the two sections. A potential difference applied across the boundary with a forward bias produces an electron flow from the n-type material to the p-type material. Forward bias means that the negative output of the supply is connected to the n-type material, and the positive is connected to the p-type, so the current flow occurs.

If the polarity of the supply is reversed there is little current flow. As explained below, photons are emitted when a current passes through a junction of this kind; the effect, known as electro-luminescence, is particularly strong for gallium arsenide crystals. The n-type impurity may be introduced into gallium arsenide by replacing some of the arsenic atoms by tellurium; the p-type may be introduced by replacing gallium atoms by zinc atoms.

If a large forward bias is applied, there is considerable recombination of electrons with the holes (an electron shortage) in the p-type material close to the junction. The recombination releases energy as photons of radiation. Recombination is confined to a very small region, about

10^{-6} m thick. When the amount of recombination with its consequent emission is high, further emission is stimulated by the first-emitted photons. This stimulated emission is in phase with the stimulating radiation and is most likely to occur in directions parallel to the plane of the junction. By polishing two opposite parallel ends of the crystal, photons may be reflected back along the boundary, so increasing the concentration of photons and producing more stimulated emission. The emission occurs in the infrared region for which gallium arsenide is transparent.

Typical dimensions for the crystal are 0·2 mm × 0·2 mm × 2·0 mm. The polished ends are chosen so that emission is in the direction of the long axis. To produce this emission high current densities are required, about 3×10^8 A/m², which corresponds to currents of the order of 100 A. These values are for room temperature operation which raises a heating problem within the crystal and usually necessitates pulsed operation. Cooling has another effect. Besides just taking away unwanted heat, it also reduces the current that is required before stimulated emission occurs. If the laser is immersed in liquid helium (temperature 4·2 K), then the necessary current drops to about 5 A. The efficiency of the conversion of electrical energy to radiation energy increases from 15 per cent at room temperature to 70 per cent at very low temperatures, which is a much higher efficiency than that of any other laser type. The outputs at these lower temperatures do not have to be pulsed and their power varies within the range of milliwatts to watts. Peak powers with pulsed semiconductor lasers may be higher, up to several hundred watts. The two main disadvantages of this type of laser are that the radiation is neither as coherent nor as highly directional as that from the other types. The semiconductor laser beam is spread through an angle of about 5° in the plane of the boundary, and about 1° at right angles to this plane.

Other semiconductor materials have been successfully used in this way, and they include indium arsenide, indium phosphide, indium antimonide and other combinations of these substances with gallium arsenide.

An advantage of semiconductor lasers is that they can be tuned to operate at a selected wavelength. This is done by altering one of the following variables: temperature, pressure or applied magnetic field.

A magnetic field is not an essential requisite for the operation of the laser, but its application enables the wavelength for maximum stimulated emission to be altered. Because of their high efficiency and easy modulation, the most successful applications of this type of laser are likely to be in the field of communications.

APPENDIX
safety

LASER power outputs range from 0·2 mW to 20 kW with a consequent change in the necessary safeguards, and the safety problems for pulsed lasers are not the same as those for continuous wave lasers. It is essential therefore to have a clear understanding of the nature of the problem and its solution before commencing laser work.

The Department of Education and Science issued an Administrative Memorandum in February 1970 entitled *Use of Lasers in Schools and other Educational Establishments*.† This is reprinted below in its entirety by kind permission of Her Majesty's Stationery Office. It should be borne in mind that experience may lead to its revision. All who use lasers should be certain that they are working to the current code of practice. Safety is the first consideration.

The first reference in the further information section of this Memorandum gives a more detailed account of all laser safety problems complete with glossary of terms, general equations for laser intensities, worked examples and list of additional references.

This Code includes a table of maximum permissible exposure levels at the cornea of the eye for different types of laser. The figure for He–Ne lasers is 3×10^{-7} W/cm² for continuous long term exposure and 3×10^{-5} W/cm² for occasional (that is accidental) exposure. The D.E.S. figure lies between these two at 10^{-6} W/cm². This is a physiological quantity of considerable significance in estimating potential danger from laser experiments.

The second reference is based on an earlier edition of the Ministry of Technology code of practice and it is a useful readable summary of that code. Those responsible for experimental work using lasers should familiarize themselves with each of these references and of course any subsequent revisions.

†A code of practice for the protection of persons exposed to laser radiation in universities is in preparation at the time of going to press.

Administrative Memorandum No. 7/70
(20 February 1970)

To Local Education Authorities
(for all secondary maintained schools),
Independent and Direct Grant Schools
(including non-maintained special schools),
Colleges of Education and Major
Establishments of Further Education

DEPARTMENT OF EDUCATION AND SCIENCE

Use of lasers in Schools and other Educational Establishments

1. Optical lasers are already in use in some schools and in some further education and teacher training establishments and it is possible that their use will become more widespread. In schools and courses of corresponding level in further education and teacher training establishments it is not expected that lasers will be used except for demonstration purposes; more advanced students may however need to carry out experiments involving the use of lasers. The purpose of this Memorandum is to draw attention to the hazards associated with their use and to give guidance on avoiding them.

2. The lasers most likely to be used for teaching purposes are low power continuous wave (CW) helium–neon lasers, which operate at a tension of up to 2 kV and currents of up to 40 mA. There is a high tension hazard, against which the usual precautions should be taken (see Appendix A of Education Pamphlet No. 53, ' Safety at School '). In addition, the powerful beams of light given out by some lasers can under certain conditions, cause damage to the body and to the eye, particularly the retina.

3. If the energy in the beam entering the eye is sufficient, local heating on the retina will kill tissue so that blindness will result, whilst higher energies will cause damage to the cornea, the iris, the lens and the eyeball itself. The subject may be quite unaware that damage is occurring, particularly with lasers operating outside the visible region. Helium-neon lasers normally operate at wavelengths of 632·8 nm but radiations of both longer and shorter wavelengths may also be emitted. For CW lasers the wavelength band that damages the retina is from 400 to 1400 nm with a peak at 570 nm.

4. For continuous-wave operation the recommended maximum for exposure of the pupil of the eye is 10^{-6} W/cm^2. Most lasers, including those used in teaching, are likely to exceed this limit, so THERE

SHOULD BE NO DIRECT VIEWING OF THE LASER BEAM UNDER ANY CIRCUMSTANCES. (Exposure time has been assumed to exceed one second in setting 10^{-6} W/cm^2.)

5. It should also be remembered that there may be reflections of the beam from specular reflectors such as mirrors, glass (including the surface of lenses), metal and glossy painted surfaces and these could reflect substantially all the energy to the eye. Furthermore, although true diffuse reflectors provide a factor of attenuation to the beam, it can and does sometimes happen that such reflectors become regular reflectors under the influence of the laser beam itself. This is particularly likely to occur when a beam has been converged by a lens.

Safety rules for the use of lasers

6. The following safety rules must therefore be observed:

   i. UNDER NO CIRCUMSTANCES VIEW THE LASER DIRECTLY. Do not use any collimating instrument such as a microscope or a telescope.

   ii. Never look along the laser beam nor expose any part of the skin to the direct beam.

   iii. Do not align a laser beam with the power on. Always use an optical alignment system first.

   iv. Check that there is no possibility of ' unsuspected ' specular reflections. Where such reflections cannot be avoided, e.g. at lens surfaces, position screens so that neither those under instruction nor the teacher will be exposed to the reflections.

   v. Position pupils or students so that screens are effective and insist that these positions are maintained.

   vi. Screens should be made of non-flammable material. They should be optically opaque and should be painted a matt grey colour.

   vii. Where lasers are being used for demonstration purposes see that no pupil or student is ever closer to any part of a laser experiment than one metre when the power is switched on.

   viii. Operate the laser in a room with as high a level of illumination as practicable, thus ensuring that the pupil of the eye is small. It is advisable, before carrying out any work with a class present, to run the laser for a short time under black-out conditions so that the presence of any stray reflections can be detected.

ix. Report any accidental exposure or even suspicion of exposure of the retina at once. If students or members of staff are working with lasers frequently, they should have their eyes examined periodically by an ophthalmologist.

x. Impress on students and pupils the danger of direct viewing or of specular reflections so that in the event of accidental exposure they will react instantly by closing the eyes and/or turning the head.

xi. Display warning notices where lasers are in use.

xii. When not in use lasers should be inaccessible to any but duly authorized members of staff.

xiii. Teachers demonstrating, and students using, lasers are advised to wear special protective goggles. These should be made of BG 18 Schotts glass 3 mm thick and are available through laboratory suppliers.

Further information

7. More detailed information can be found in the 1969 *Guide on Protection of Personnel against Hazards from Laser Radiation.* BS 4803 : 1972. British Standards Institution, 2, Park Street, London, W1A 2BS, and *A General Guide to the Safe Use of Lasers*, obtainable from the Electronic Engineering Association, Berkeley Square House, Berkeley Square, W1 X6JU.

8. Users of pulsed ruby lasers should refer to the ' General Guide ' which is itself based on the ' Code of Practice '.

INDEX

Aberration 55
  spherical **55–59,** 63
Absorption 77, 78
  of photons 76
Airy disc 5, 6
Ammonia maser 107–108
Amplification of photons 82
Amplitude 14
  division of 19, 22
Aperture, construction of 4, 5
Aquadag 10
Argon-ion laser 118
Astigmatism 55, **61–62,** 63
Atom, excited 80–84

Beam (laser) divergence 45
Beam (laser) expansion **8,** 11, 55, 56
  checking parallel nature 9
Biprism, Fresnel, complex experiment 15–18
  simple experiment 13–14
Bohr 75
Boltzmann (Maxwell) distribution 82–84
Brewster angle 66–69
  window **66–67,** 117, 120

Carbon dioxide laser 120–121
Cavity, resonant **101,** 104, 106, 108, 109
Circle of least confusion 56
Circular aperture, diffraction by 4, 5, 6
Coherence **86–97,** 111
  length 90, 96–97
  length of lasers 97
  of laser 1, 10, 21, 58
  time 90
Coma 55, **60–61,** 63
Complex diffraction patterns 10
Continuous wave laser 111–112
Contour fringes 22
Criterion, Rayleigh's 8, 9

Dark room requirements for holography 48
Davisson 74, **79**
D.C. excitation 118
De Broglie 74
Depth of focus 54
Design angle 47
Dextro-rotatory 73
Diffraction by circular aperture 4, 5, 6
  by dust particles 10, 54, 58
  by rectangular aperture 4
  by single slit (Fraunhofer) **1–3,** 13, 16, 22, 23
  (Fresnel) **22–23,** 24
  by square aperture 4
  by straight edge **13,** 18, 24
Diffraction, Fraunhofer 19, 23
  Fresnel 19, 23–27
  grating **27–31,** 40–41
  basic equation 27
  demonstration with laser 28
  metal rule experiment 30
  non-normal incidence 28, 29
  reflected orders 28
  reflection grating experiment 30
  spacing 29
Diffraction, influence on aberration 58, 60, 61
  interference patterns 14, 15–18, 19
Discharge lamp 1
  for viewing holograms 49
Distortion 55, **62–63**
Distribution, Maxwell-Boltzmann 82–84
Divergence of laser beam 45
Division of amplitude 19, 22
  wavefront 19, 22
Doppler effect 69, 105, 106
Double refraction 71–72
Double-slit interference patterns 10, 11, **12–13,** 15, 16, **86–89**

Electro-luminescence 121
Electromagnetic wave 64–65

Electron   7, 76, 79
  energy levels   75, 76, 79
  wave properties   74, 79
Elliptical reflection tube   113–114
Emission, photon   76, 91, 121
  spontaneous   80
  stimulated   80–82, 84
Energy bands   79, 80, 92
  levels   75, 76, 78, 79
Energy levels, electron 75, 76, 79
  in helium and neon atoms   116
  in hydrogen atom   75, 76
  range of   79
Energy state   78, 79
Equal amplitude condition   14
Excitation by direct current   118
Excited atom   80–84
  state   80–84, 107
Exclusion principle (Pauli)   78
Expansion of laser beam   **8**, 11, 55, 56
Extraordinary ray   71–72

Fabry–Pérot resonator   102
Field curvature   55, **62–63**
Focus, depth of   54
  Gaussian   58
Focusing, hologram experiment 42–45
'Forbidden' transitions   81, 110
Fourier synthesis   92, 96
Fraunhofer diffraction, definition **19**, 23
  patterns   1–3, 13, 16, 22, 23
Fresnel biprism, complex experiment 15–18
  simple experiment   13–14
Fresnel diffraction, definition   **19**, 23
  patterns   22–27
  zones   31–34
Fringes, contour   22

Gallium arsenide laser   121–122
Gas lasers   114–121
Gaussian focus   58
Germer   74, **79**
Glass (with neodymium) laser   114
Ground state   80

Harmonics   100–101
Heisenberg's uncertainty principle   78, 79, **90–92**, 94
Helium-neon laser   47, 81, 105, **114–118**
  Brewster windows   117
  construction of   115, 117–118
  d.c. excitation of   118
  energy levels   116

Hologram   35
  broken   54
  depth of focus   42–45
  transmission   45–46
  viewing of   48–50
Holography, dark room requirements for   48
  films for   47, 48
  multiple image   53–54
  objects for   46, 51
  path difference for   47, 52
  photographic plates for   47, 48
  principles of   35–41
  reflection   51–53
  self   53
  stability for   45, 46, 50, 51
  three-dimensional images from   41–45
  transmission   45–47
  uni-phase laser action   105
Hydrogen atom, energy levels   75, 76
  spectral series   76
  spectrum   77
Huyghens' principle of secondary wavelets   29, 30, 86–88, 92

Interference mirror   104
  multiple   21
  of wave packets   96
  pattern, double-slit   86–89
    Fresnel biprism   13–18
    hologram   40
    influence of diffraction on   14, **15–18**, 19
    Lloyd's single mirror   11–13
    number of photons in   88–89
    Young's slits   **10–11**, 14, 15, 86–89
Inverse square law experiment   68–69
Ion lasers   118–119

Kerr cell   112
Krypton-ion laser   118

Laevo-rotatory   73
Laser   84
  action   106
  alignment of mirrors   104
  argon–ion   118
  beam, divergence of   45
    expansion   **8**, 11, 55, 56
    expansion, checking parallel nature   9
  multimode   102–103, 104–105, 106
  carbon dioxide   120–121

coherence 1, 10, 21, 58
  length 90, 96–97
  time 90
continuous wave 111–112
gallium arsenide 121–122
glass (with neodymium) 114
helium-neon
  see helium-neon laser
krypton-ion 118
line width 105–106
neodymium in glass 114
photons in pulse of 111
ruby 108–112
transitions 110–111, 115–116
Lasers, gas 114–121
  ion 118–119
  pulsed 111, 112, 113
  semiconductor 121–122
  solid state 114
Lattice vibrations 80
Law of Malus 70–71
Length of wave packet 95–97
Lens aberrations 55
Lifetime 80
Light meter 14, 68
Limit of resolution 6–10
  definition for microscope 7
  definition for telescope 8
  measurement of 8–10
Line width 97
  due to stimulated emmision 105
  laser 105–106
  mechanical demonstration of 98–100
  resonator 105–106
Littrow prism 119–120
Lloyd's single mirror 11–13

Magnification by projection 20, 22, 56, 57
Maiman 108
Malus's law 70–71
Maser 107–108
Maxwell 74
Maxwell–Boltzmann distribution 82–84
Measurement of wavelength 2, 21–22, 29–30
Mechanics, wave 78
Metal rule experiment 30
Metastable state 109–110
Microscope objective 9, 45, 57
Mirror, alignment for laser action 104
  interference 104
  Lloyd's single 11–13

Multimode laser beam 102–103, 104–105, 106
Multiple image holography 53–54

Newton **19–22,** 74
Newton's rings 19–22
  experiment with laser 19–20
  phase change 20
  projection of 20
Neodymium in glass laser 114

Optical activity 73
Optical flat 20, 21
Ordinary ray 71–72

Path difference, holography 47, 52
  large 21
  wave packet 94–95
Pauli exclusion principle 78
Phase change with Newton's rings 20
Photocell 68, 69, 70
Photograph, lack of three dimensions 43, 45
Photographic film for holography 47, 48
  plates for holography 47, 48
  reproduction of apertures 5
Photon 74, 76, 81
  absorption 76
  amplification of 82
  emission 76, 91, 121
  in a laser pulse 111
Photons in an interference pattern 88–89

Physical optics experiments, general 1, 104
Planck 74, 75, 79
Planck's constant 75, 79
Plane of vibration 69, 72, 80
Plane polarized wave 64, 65, 70, 71, 73
Polarized light, scattering 65–66
Polarized plane wave 64, 65, 70, 71, 73
Polaroid 64–68, 70, 73
Population inversion 82–84, 110
Population normal 83
Pressure broadening 96, 105
Principle of exclusion (Pauli) 78
  of superposition 37, 39, 92
  of Uncertainty (Heisenberg) 78, 79, **90–92,** 94
Principles of holography 35–41

Prism, Littrow 119–120
  small angle 46
Probability 79, 94, 110
Probable distribution 91–92
Projection for magnification 20, 22, 56, 57
  Newton's rings 20
Pulsed lasers 111, 112, 113

Q-factor 100
Q-switching 112–113
Quanta 74
Quantum theory 74

Rayleigh criterion 8, 9
Rectangular aperture, diffraction 4
Reference beam 36–37, 38, 39, 51, 52
Reflection grating experiment 30
  holography 51–53
Refraction, double 71–72
Resolution, limit of 6–10
  definition for microscope 7
  telescope 8
  measurement of 8–10
Resonant cavity **101,** 104, 106, 108, 118
Resonator, Fabry–Pérot 102
  line width 105–106
Reversal of sodium D lines 77, 104, 106
Ruby crystal 108–109
  laser 101–112
Rule, metal, experiment 30
Rutherford 75

Safety **123–126**
  in forward projection 56
  in viewing holograms 48–50
Scattering 65–66
Schrödinger's wave equation 78, 79
Secondary wavelets 29, 30, 86–88, 92
Selection rules 81, 110
Self-holography 53
Semiconductor lasers 121–122
Signal beam 36–37, 38, 39
Single mirror, Lloyd's 11–13
Single slit diffraction, Fraunhofer **1–3,** 13, 16, 22
  Fresnel 22–23, 24
Smoke box 3, 4, 29, 61
Sodium D lines 77, 104, 106
Sodium flame experiment 76–78

Solid state lasers 114
Spectrum, continuous 77
  hydrogen 77
  white light 77
Spherical aberration **55–59,** 63
Spontaneous emission 80
Square aperture, diffraction 4
State, excited 80–84
  ground 80
  metastable 109, 110
  stationary 79, 80
Stationary state 79, 80
Stationary wave 85, 98
Stimulated emission **80–82,** 84
  line width 105
Straight edge diffraction **13,** 18, 24
Superposition, principle of 37, 39, 92
Synthesis, Fourier 92, 96

Thomson G. P. 74
Three-dimensional images, holography 41–45
Transitions, 'forbidden' 81, 110
  laser 110–111, 115–116
Transmission holograms 45–47

Uncertainty principle (Heisenberg) 78, 79, **90–92,** 94
Uni-phase 102
Unpolarized light 65, 66, 71

Vibrations, lattice 80
Viewing of holograms 48–50

Wave, electromagnetic 64–65
  equation, Schrodinger 78, 79
Wavefront, division of 19, 22
Wavelength, measurement of 2, 21–22, 29–30
Wave mechanics 78
Wave packet 74, 93, 96
  length 95–96
Wave particle duality 74, 89–90
  properties of an electron 74, 79
  stationary 85, 98
Window, Brewster **66–67,** 117, 120

Xenon flash tube 108–109

Zone plate 32–34
Zones, Fresnel 31–34

THE WYKEHAM SCIENCE SERIES

1. *Elementary Science of Metals* — J. W. MARTIN and R. A. HULL
2. *Neutron Physics* — G. E. BACON and G. R. NOAKES
3. *Essentials of Meteorology* — D. H. MCINTOSH, A. S. THOM and V. T. SAUNDERS
   (Paper and Cloth Editions available)
4. *Nuclear Fusion* — H. R. HULME and A. MCB. COLLIEU
5. *Water Waves* — N. F. BARBER and G. GHEY
6. *Gravity and the Earth* — A. H. COOK and V. T. SAUNDERS
7. *Relativity and High Energy Physics* — W. G. V. ROSSER and R. K. MCCULLOCH
8. *The Method of Science* — R. HARRÉ and D. G. F. EASTWOOD
9. *Introduction to Polymer Science* — L. R. G. TRELOAR and W. F. ARCHENHOLD
10. *The Stars; their structure and evolution* — R. J. TAYLER and A. S. EVEREST
    (Paper and Cloth Editions available)
11. *Superconductivity* — A. W. B. TAYLOR and G. R. NOAKES
12. *Neutrinos* — G. M. LEWIS and G. A. WHEATLEY
13. *Crystals and X-rays* — H. S. LIPSON and R. M. LEE
14. *Biological Effects of Radiation* — J. E. COGGLE and G. R. NOAKES
    (Paper and Cloth Editions available)
15. *Units and Standards for Electromagnetism* — P. VIGOUREUX and R. A. R. TRICKER
16. *The Inert Gases: Model Systems for Science* — B. L. SMITH and J. P. WEBB
17. *Thin Films* — K. D. LEAVER, B. N. CHAPMAN and H. T. RICHARDS
18. *Elementary Experiments with Lasers* — G. WRIGHT and G. FOXCROFT
19. *Production, Pollution, Protection* — W. B. YAPP and M. I. SMITH
    (Paper and Cloth Editions available)
20. *Solid State Electronic Devices* — D. V. MORGAN, M. J. HOWES and J. SUTCLIFFE
21. *Strong Materials* — J. W. MARTIN and R. A. HULL
22. *Elementary Quantum Mechanics* — SIR NEVILL MOTT
    (Paper and Cloth Editions available)
23. *The Origin of the Chemical Elements* — R. J. TAYLER and A. S. EVEREST
24. *The Physical Properties of Glass* — D. G. HOLLOWAY and D. A. TAWNEY
25. *Amphibians* — J. F. D. FRAZER and O. H. FRAZER
26. *Chemical Engineering in Practice* — G. NONHEBEL and M. BERRY
    (Paper and Cloth Editions available)
27. *Temperature Regulation* — S. A. RICHARDS and P. S. FIELDEN
    (Paper and Cloth Editions available)

THE WYKEHAM TECHNOLOGY SERIES

1. *Frequency Conversion* — J. THOMSON, W. E. TURK and M. J. BEESLEY
2. *Electrical Measuring Instruments* — E. HANDSCOMBE
3. *Industrial Radiology Techniques* — R. HALMSHAW
4. *Understanding and Measuring Vibrations* — R. H. WALLACE

All orders and requests for inspection copies should be sent to the appropriate agents. A list of agents and their territories is given on the verso of the title page of this book.

75560